BONE DETECTIVE

WOMEN'S ADVENTURES IN SCIENCE

BONE DETECTIVE

the story of forensic anthropologist

DIANE FRANCE

by Lorraine Jean Hopping

Franklin Watts
A Division of Scholastic Inc.
New York • Toronto • London • Auckland • Sydney
Mexico City • New Delhi • Hong Kong
Danbury, Connecticut

Joseph Henry Press
Washington, D.C.

Author's Acknowledgments

Thank you, Diane France, for educating me about the beauty of bones and for entertaining me with your true-life adventures. This book owes a lot to your patience, openness, and generosity. Thanks also to Dave and Dolores France for sharing your home and your daughter, and to Joy Metzger and LuAnn Schellhaas for sharing your childhood memories. Many other busy people kindly agreed to interviews and helped enrich this book: Tom Adair, Jane Bock, Tom Crist, Tony Falsetti, Jennifer Fillion, Mike Gear, Cal Jennings, John Lindemann, Lynn Kilgore, Ed Killam, Julie Kovats, John McPhail, Ed Pearl, Jim Reed, Peter Sarandinaki, Frank Saul, Julie Saul, Paul Sledzik, Jack Swanburg, Vickey Trammell, Cecilia Travis, Mary Wright, and several NecroSearch members to whom I didn't get a chance to speak directly but to whom I listened with open ears at the meeting. Finally, I'm especially grateful to my husband, Chris Egan, who offered support that was both logistical and incalculable.—LJH

Cover photo: Diane France measures and examines human bones in her lab. Remembering that they are the remains of what was once a living person, Diane handles the bones respectfully.

Cover design: Michele de la Menardiere

Library of Congress Cataloging-in-Publication Data

Hopping, Lorraine Jean.
Bone detective : the story of forensic anthropologist Diane France / Lorraine Jean Hopping.
p. cm. — (Women's adventures in science)
Includes bibliographical references and index.
ISBN 0-531-16776-3 (lib. bdg.) 0-309-09550-6 (trade pbk.) 0-531-16951-0 (classroom pbk.)
1. France, Diane L. 2. Women forensic anthropologists—United States—Biography—Juvenile literature. I. Title. II. Series.

GN50.6.F8H66 2005
363.25'092—dc22

2005000784

Printed in Mexico.
2 3 4 5 6 7 8 9 10 R 14 13 12 11 10 09 08 07 06

About the Series

The stories in the *Women's Adventures in Science* series are about real women and the scientific careers they pursue so passionately. Some of these women knew at a very young age that they wanted to become scientists. Others realized it much later. Some of the scientists described in this series had to overcome major personal or societal obstacles on the way to establishing their careers. Others followed a simpler and more congenial path. Despite their very different backgrounds and life stories, these remarkable women all share one important belief: the work they do is important and it can make the world a better place.

Unlike many other biography series, *Women's Adventures in Science* chronicles the lives of contemporary, working scientists. Each of the women profiled in the series participated in her book's creation by sharing important details about her life, providing personal photographs to help illustrate the story, making family, friends, and colleagues available for interviews, and explaining her scientific specialty in ways that will inform and engage young readers.

This series would not have been possible without the generous assistance of Sara Lee Schupf and the National Academy of Sciences, an individual and an organization united in the belief that the pursuit of science is crucial to our understanding of how the world works and in the recognition that women must play a central role in all areas of science. They hope that *Women's Adventures in Science* will entertain and enlighten readers with stories of intellectually curious girls who became determined and innovative scientists dedicated to the quest for new knowledge. They also hope the stories will inspire young people with talent and energy to consider similar pursuits. The challenges of a scientific career are great but the rewards can be even greater.

Other Books in the Series

Beyond Jupiter: The Story of Planetary Astronomer Heidi Hammel

Forecast Earth: The Story of Climate Scientist Inez Fung

Gene Hunter: The Story of Neuropsychologist Nancy Wexler

Gorilla Mountain: The Story of Wildlife Biologist Amy Vedder

Nature's Machines: The Story of Biomechanist Mimi Koehl

People Person: The Story of Sociologist Marta Tienda

Robo World: The Story of Robot Designer Cynthia Breazeal

Space Rocks: The Story of Planetary Geologist Adriana Ocampo

Strong Force: The Story of Physicist Shirley Ann Jackson

Contents

For the Love of Bones

Diane France loves bones. Does that sound weird? It will make perfect sense in a moment, as soon as you read her amazing story. Diane is a forensic anthropologist, a bone detective. Skeletons are her key to a life full of grand adventures.

Bones lead Diane all around the world and back in time. One day she's in Russia "meeting" the skeletons of a royal family who lived a century ago. Another day she is peering into the empty eye sockets of an American outlaw. Meanwhile, the skulls of eight Civil War soldiers line the shelves of her Colorado lab.

Diane is thrilled, most of all, when she uses her science know-how to help people struck by disaster. Moments after a phone call, she rushes to the scene of a plane crash, a fire, an accident, or terrorist bombing to identify victims. She examines the bodies of murder victims for clues about how they died. When a body is missing, she searches for it with a team that specializes in finding hidden graves.

Why does Diane France love bones? How did she become a bone detective? This small-town girl let science lead the way to a world of adventure.

Why does Diane France

love bones?

How did she become

a bone detective?

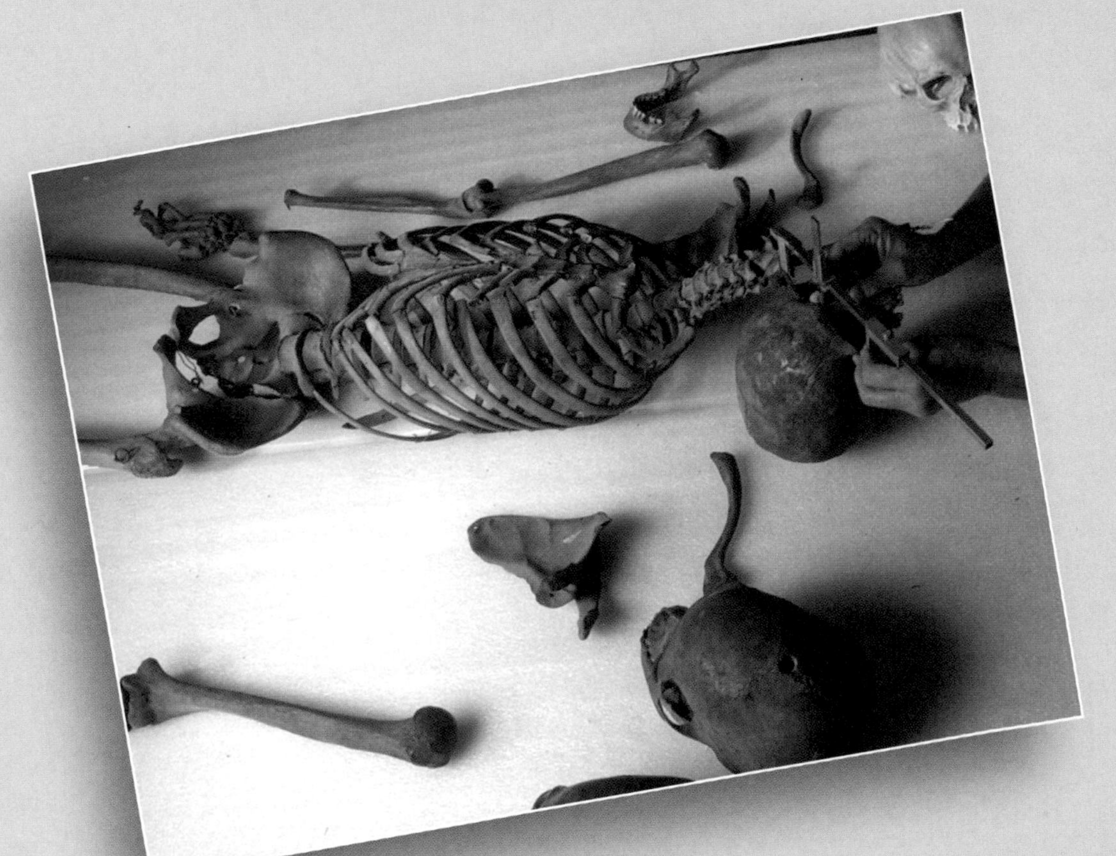

This small-town girl

let science

lead the way to

a world of adventure.

1

Diane France's Brain

A brain is floating in a plastic bucket that's sitting, alone, on a table. The hand-printed label says "Diane France's brain."

The people who file in and out of the room know that name well. Dr. Diane France is a well-respected scientist—and has been for nearly 20 years. To many of these people, she is also a friend.

A visitor blurts out, "Oh no! What happened to poor Diane?"

There's a moment of silence as eyes shift from the visitor to the bucket and back to the visitor. Then, everyone cracks up. Between chuckles, someone explains that the brain does belong to Diane France; that's true. But it isn't her brain, the one inside her head. That brain is still alive and thinking and, at this very moment, sharing its knowledge with other scientists just a couple of miles away.

Diane is in Washington, D.C., to teach a course at the 1994 Armed Forces Institute of Pathology (AFIP) annual conference on anthropology. For now her mind is on bones, not brains. She is teaching participants how to tell the life story of a dead person by examining the skeleton in fine detail. After the class she plans to drop off her bone specimens and pick up her brain at the AFIP's museum, the National Museum of Health and Medicine (NMHM). The brain—the one in the plastic bucket—is one of 10,000 anatomical specimens housed in the museum.

Diane France handles bones almost every day, measuring them *(opposite)* and studying them for clues about a person's life and death *(above)*. She reminds herself to "handle with care" out of respect for the original owner: a living human being.

Paul Sledzik *(right)* investigates the contents of a 17th-century grave for a study of life in colonial Maryland. Diane France *(below)* creates a fake crime scene for a class she teaches on recovering evidence.

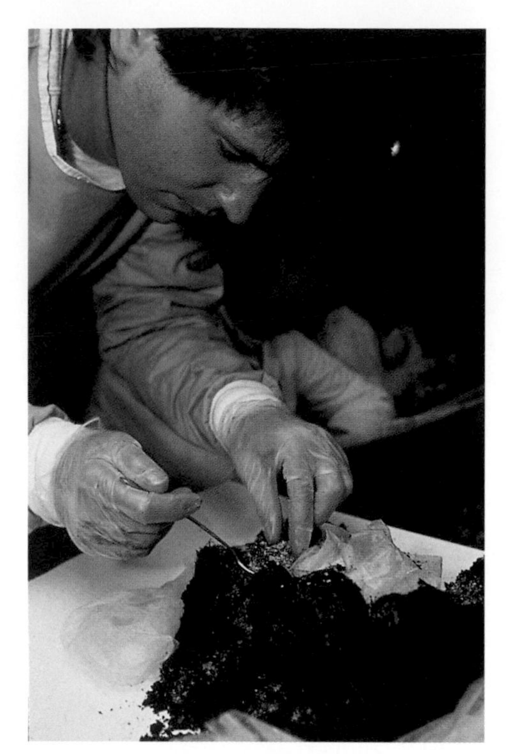

A buddy of Diane's, Paul Sledzik, is in charge of these brains, hearts, lungs, limbs, bones, samples of blood and skin tissue, and other preserved body parts. Paul happens to be an expert on vampire legends, but that's not what earned him this macabre job. He's a physical anthropologist, a scientist skilled in identifying skeletal remains.

So is Diane. That's why, when she asked Paul for a brain one day, she knew he wouldn't blink.

What Is Forensic Anthropology?

Anthropology comes from the Greek word for the "study of human beings." We humans are a complex bunch, so there are four areas of specialty:

cultural anthropology (societies—past and present)

linguistic anthropology (languages)

archaeological anthropology (past civilizations studied through their artifacts)

physical anthropology (all variations in the biology of the body, including human ancestors and relatives)

Forensic anthropologists are physical anthropologists who examine human remains that are part of an investigation—a murder, an accident, or a disaster, for example.

~ *Hand Me a Tissue*

Anthropologists routinely handle "specimens" or "anatomical samples" or "biological material." Sometimes, they just call it "tissue." All of these shapeless terms veil the human truth—that every body part once belonged to a living being. Diane believes that this fact is very important to remember and respect. It's something she tells her students as soon as she introduces them to human bones. On the other hand, working with body parts is creepy, even revolting at times. Using scientific terms can make a rather nauseating job easier to handle.

Humor helps, too. The "Diane France's brain" story is already making the museum rounds. In fact, this "just another day at the office" attitude is why Paul labeled the brain in the first place. He didn't want a well-meaning curator (a museum worker) to reshelve the brain before Diane had a chance to pick it up.

Diane is collecting brains as a job, not a hobby. The National Zoo in Washington, D.C., hired her to make casts—or detailed copies—of animal brains for an exhibit. So far she has rounded up a nut-sized squirrel brain, an orangutan brain the size of a large fist, and a fin whale brain the size of a basketball. These specimens are waiting for her at the Smithsonian Institution in the heart of Washington, D.C. Later, she will fly to Detroit, Michigan, to borrow an elephant brain from a biologist. "Diane France's brain"—the one at the museum—is human.

Scientists study body parts at the NMHM in Washington, D.C., to learn about injuries, diseases, and medical treatments from Civil War times to the present.

Many of the specimens at the NMHM are historic. The museum began collecting them in the 1860s, during and after the American Civil War. But Paul made sure that Diane's brain is "pristine and modern" so that she can produce a good cast. The well-preserved brain is from a recently deceased cadaver, a body donated for use in medical schools and research.

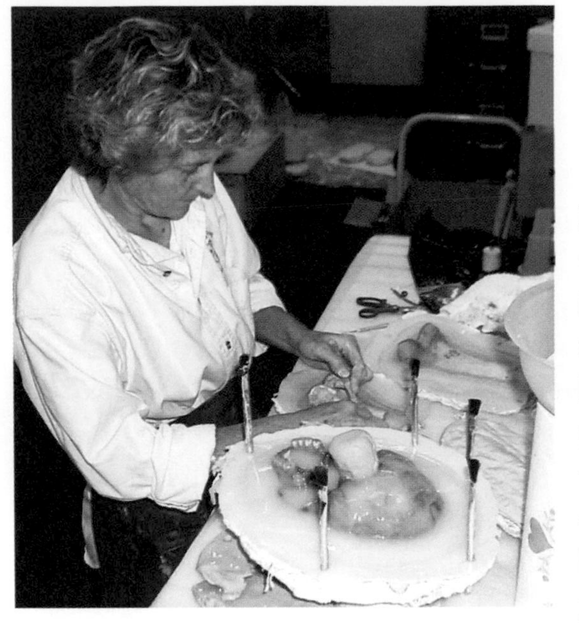

Diane makes a rubber mold of a cranium, the top part of the skull. Liquid plastic poured into the mold will harden quickly into a cast—a high-quality copy of the cranium. Diane's casts sell to zoos, museums, schools, and anyone else who wants to bone up on anatomy.

Finding fresh brains has turned out to be the easy part. Making brain casts will be a real intellectual challenge. Diane usually makes plastic casts of hard tissue—bones, teeth, and cartilage—but brains are floppy and jiggly. Set one on a table and it droops like an underfilled water balloon. Somehow, Diane will have to make these brains keep their true shape while she makes a mold. There's no "how-to" book on the subject, so she'll have to use her own brain to invent a way.

~ *What a Day to Dress Up*

After Diane's class ends and the AFIP conference winds down, she packs up her bones and gets ready to swing by the NMHM. She opens the door to a sky dumping buckets of rain. She is seriously dressed up in a pink silk blouse, black suit jacket, matching black slacks, high heels, dangling earrings, and a long gold necklace. She has no umbrella. She has no car. Luckily, she has another good friend.

Tom Crist is an archaeologist, someone who normally hangs around artifacts that are old and decrepit. But when it comes to cars, he prefers one that is new and shiny. He offers to drive Diane in his spotless Pontiac Bonneville—"dark green with tan leather seats," he adds.

It's a no-brainer. Minutes later, at the museum entrance, Diane unloads her box of bones in the downpour. The rain shows no sign of easing up, so Tom kindly agrees to drive Diane downtown, to the Smithsonian Institution, where she plans to make her casts. Diane asks him to wait—it will take just a minute—while she collects her brain.

She winds her way through the museum to a small back room and spots "Diane France's brain" on a table. She picks up the plastic, two-gallon bucket. It's light—about the weight of half a gallon of milk—even with the liquid preservative. Diane recognizes the odor of formalin: sharp with a hint of wintergreen, but not as pungent as formaldehyde, the preservative in its pure form. In fact, it's barely noticeable through the bucket's plastic lid.

~ *Not-So-Easy Does It*

As Diane heads back to the museum lobby, she tries to keep the brain from sloshing around too much. The high heels sure don't help. Wearing them, or any shoes, for that matter, isn't her usual style. She and her malamute dog, Moki, both pad around barefoot in her casting lab in Fort Collins, Colorado. It's one of the perks of owning the place, and her sole human employee doesn't mind a bit. Diane even gives lectures and teaches workshops while barefoot, a habit she started in order to feel less nervous and now does for comfort. Sometimes, she invites the audience to take off their shoes, which gets a few laughs and puts everyone at ease.

Bucket firmly in hand, high heels balanced, Diane walks out the door of the small specimen room. She goes around the bend, past some filing cabinets and shelves, down a long, narrow passageway, a turn to the left, and, the tricky part: a one-handed push through a brass-and-glass door. Then, both hands back on the bucket, she continues down the long corridor to the lobby, where her heels echo—click, click, click—across the marble floor.

The lobby is a public exhibit area. This being a museum of preserved body parts, the displays both repulse and fascinate visitors.

Diane's pet malamute, Moki, knows not to stick her nose into anything in Diane's lab. She's content to sit and watch the bones go by.

Diane doesn't even notice what's in the glass cases. She's focused solely on the specimen in her hands. Through the exit doors, she can see that the rain hasn't let up and that Tom's car is parked in a driveway about 15 feet away. She'll have to make a run for it—high heels, brain, and all. Paul opens one of the museum's four giant doors, and Diane steps out onto a large roofed terrace. Grasping the bucket with both hands, she makes a beeline for the Pontiac.

Tom has opened the back door so that she can slide right in. She plops down on the smooth leather seat and, in the blink of an eye, the unthinkable happens: The plastic bucket flexes in her hands, the lid pops off, and the brain splashes out on her lap. Without thinking, Diane scoops it up and pops it right back in the bucket.

Then, she thinks, *Wow, this burns*.

~ Feeling the Burn

Formalin is all over Diane's lap, her hands, and—*oh man, poor car!* Mortified, she notices that the fine leather in Tom's brand new ride has taken a direct hit. She feels bad about the car—she really does—but her legs and hands feel like they're on fire. Diane jumps out into the downpour, hoping the rain will wash away some of the formalin.

"Oh man, it's still burning," Diane says, waving raindrops onto her body. Formalin can sting the nostrils a little if you sniff it up close. Diane begins to wonder what it might be doing to her suit pants, her silk blouse, her skin, and—*oh no, Tom's new leather seats!* Her skin wins out.

Diane runs to the museum's ladies' room. She strips off the black pants . . . the suit jacket . . . that fancy pink blouse . . . and splashes handfuls of water on her stinging skin. Finally, after some wet paper towel action, the pain eases to a tolerable level. But then, Diane realizes, she's standing there in her underwear. She puts on her blouse and high heels and looks at the formalin-soaked slacks on the floor. *No way am I wearing those things anytime soon.*

There's a knock on the restroom door and then a familiar voice.

"Are you okay, Diane?" her pal Paul asks from the other side.

She calls back that she's more embarrassed than hurt. "Would you happen to have anything I could wear?"

Moments later an arm appears through the door holding a pair of men's gym shorts. Paul's not a large man, but Diane is petite—short and thin with a small frame. She puts on his shorts, but the baggy things won't stay up. She'll have to hold them. She pinches the waistline tight in her fist and walks awkwardly out of the ladies' room, still in her high heels.

The plastic bucket flexes in her hands, the lid pops off, and the brain splashes out on her lap. Without thinking, Diane scoops it up and pops it right back in the bucket.

Paul can't help it. He starts laughing, which gets Diane laughing too. She does look ridiculous, and these guys are going to tease her about it forever, but at least she's okay. She wonders if Tom can say the same about his car.

Still holding up the shorts, she dashes out the door and through the rain. Diane looks first inside the bucket on the floor of the car. No brain damage, she's relieved to see. There's even enough formalin left to cover the tissue. Then, she glances at Tom's leather seat. It's both soggy and stinky. It will eventually dry, but Diane can only hope the stench of formalin will fade with time. Tom tells her not to worry; it's just a car. *A brand new Pontiac Bonneville, dark green with tan leather seats,* Diane can't help thinking.

~ *A Drop in the Bucket*

Clearly, Diane has no choice but to call it a day, so Tom drops her off at her hotel. Diane's mother, Dolores, has come along for the trip and watches, wide-eyed, as her 40-year-old daughter walks into their room. Diane is carrying a brain in a bucket in one hand, holding up baggy gym shorts with the other, and reeking of an odor not found in nature. She's still wearing the heels, earrings, and gold necklace, but now they look comically out of place.

Dolores can't imagine what has happened, but the sight before her is not a total surprise. Here's her only daughter, a well-respected scientist with a Ph.D., looking like the unruly tomboy she once was. Dolores remembers dressing Diane in ladylike outfits and pin curls, only to find her playing in a mud puddle moments later. Mother and daughter share a good laugh as Diane heads for a hot shower.

The next day Diane carries the brain to the Smithsonian on the Metro train without mishap, though she has no trouble imagining a spill—kerplunk, kerplunk, kerplunk—on the long escalator ride down to the subway. As the Smithsonian curators hand over their squirrel, orangutan, and fin whale brains, they mention they're worried about damage, especially to the whale brain. At about 15 pounds, that brain takes two very strong hands just to hold it. Diane assures them that she's a trained professional—she knows what she's doing.

Then she thinks, *But I still don't know how to do it. How in the world am I going to make a mold of these floppy brains?*

That hurdle turns out to be a lot higher than Diane expects. In one of the Smithsonian's curation rooms, she carefully places "Diane France's brain" on a table and props it up around the edges with clay. But the clay doesn't stick to the table, and the brain loses its shape. That's no good.

"Diane France's brain" made it from plastic bucket to bronze cast with all its folds, grooves, and other details intact.

Next, she tries putting the brain in a pan and smooshes clay around the inside edge of the container. Still no good.

Finally, after more trial and error, Diane grabs a stainless steel bowl—just like one in her kitchen—and presses clay in a ring around the curved bottom. She adds a layer of formalin-soaked paper towels to keep the bottom of the brain damp. Then, she carefully nestles the brain in the ring. It works! The brain holds its shape as she coats the top with a liquid form of silicone rubber. The rubber hardens to make one-half of a mold. To make the other half, all she'll have to do is flip over the brain. *Mission accomplished.*

Please touch: A girl examines a bronze cast of a human brain at the Think Tank exhibit at Washington's National Zoo, while another visitor probes a bronze cast of an elephant brain.

Back in Colorado, Diane uses the brain molds to make bronze casts for the National Zoo's Think Tank exhibit. They look so cool she makes two extra copies. She gives one to her father, Dave. He's a doctor, and they share a love for anything anatomical. Diane keeps the other brain for herself as a reminder of her latest adventure. It's just a drop in the bucket compared to her many other "great moments" in science, but the memory will always make her chuckle. Humor helps lighten the darker side of her life as a forensic anthropologist, as she analyzes human remains that are evidence in an investigation—murder victims, for instance.

Both casts and crimes have taken Diane around the world, but the journey all began in a tiny town called Walden. There, as a young girl, Diane France's brain was full of curiosity—and a little mischief, too.

The journey all began in a
tiny town
called Walden.

There, as a young girl,
Diane France's **brain** *was full of* **curiosity**
—and a little mischief, too.

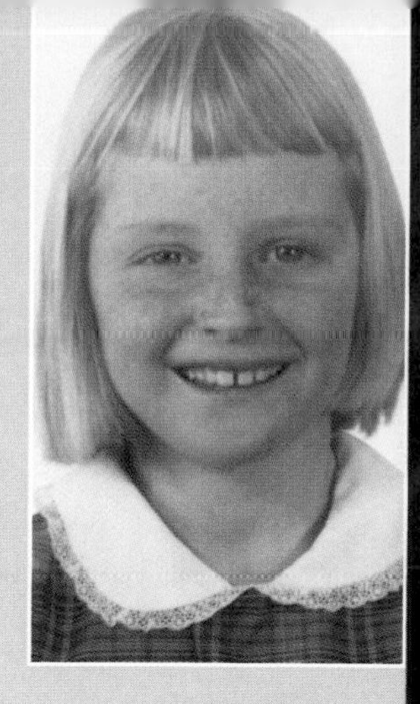

2

The Doctor's Daughter

One cold October day, Diane watched Dad dress up his skeleton. She loved that skeleton. Those plastic bones, wired together at every joint, had hung in her father's office for as long as she could remember.

Dad normally used the skeleton as a model to show his patients what their bones looked like. But today was Halloween, so he put a cigar in its teeth, a knife in one bony hand, a pumpkin in the other and—*Boo!* He had a scary decoration, eerily lit from below, for his office window. He even taped a picture of bad lungs on its chest to suggest it had smoked too much.

Not to be outdone by a bag of bones, Dad dressed himself up in a bright red devil suit that Mom had made. It had a mask, a hood with two horns, and a long curly tail. Everyone in town thought Dad and Mom's Halloween creations were a real hoot.

"Everyone" was no exaggeration. As of 1960, Walden, Colorado, had about 800 people, and they all knew Dr. and Mrs. France. Dave, as everyone called him, was the only doctor in a wide stretch of Rocky Mountain wilderness. He set all the broken bones, stitched all the cuts, took out tonsils, and treated every other illness or injury that came through his door. Day or night, Diane's dad delivered babies—including most of her friends.

Growing up, Diane's white shoes, spotless dresses, and neatly combed bangs stayed that way about as long as a camera click. She loved to play in the mud from the time she could walk (*shown opposite with her father*).

When someone died, he switched hats from doctor to coroner and took care of the body.

If the doctor had a specialty, it was removing fish hooks—both from tasty fish and from clumsy fishermen. Walden was one of the top fishing spots in the country.

~ *A Devil at the Wheel*

As the only doctor in town, Dad was on call for emergencies around the clock. Often there would be a knock on the door or a phone call, and Dad would disappear into the night with his medical bag. That Halloween evening was no exception; Dad had to leave in the middle of a costume party.

You can still watch *Bonanza*, Diane's favorite TV show as a child, in reruns. But do you have an autographed picture of the cast, like Diane's?

This time a man had a perforated ulcer, a hole in his intestine, and needed immediate surgery. The nearest hospital was in Laramie, Wyoming, more than an hour's drive away. Walden had no ambulance, but the post office had a station wagon. Still in his red devil suit—there was no time to change clothes—the doctor loaded his patient into the postal vehicle and gave him a painkiller.

The man had also been drinking, and both the medicine and the alcohol made him drowsy. During the long drive, he would start to wake up, see a red devil at the wheel, and then pass out again. By the time they reached Laramie the man believed he was in a very bad place. "Satan" was standing over him, staring straight into his eyes. The patient sobered up fast—just in time for a devil of a surgery.

Dad told the "devil at the wheel" story again and again, each time a little funnier. Diane laughed at all his stories. They were the Wild West adventures of a country doctor. Walden even looked like the western town in *Bonanza*, Diane's favorite TV show. The whole family watched that show every Sunday evening with soda pops and a bowl of popcorn. Diane treasured her autographed photo of the cast.

~ *Picture Perfect*

Walden's Main Street—all two blocks of it—had the Fair Share grocery store, Shawver Drugs, the Walden Cafe, the Park Theater ("25-cent movies every Wednesday night!"), the Sportsman Supply store, a hotel, a couple of bars, and a gas station. The street was capped off by an impressive county courthouse with pillars.

But all things considered, the town was just a speck on a vast, wall-to-wall carpet of sagebrush and grass. Golden hayfields rolled far into the distance in every direction, stopped only by a rim of gray mountain ridges with names like Rabbit Ears and Medicine Bow.

Diane's hometown of Walden, Colorado *(above)*, had a population of about 800 when she was growing up there in the late 1950s. Diane appears in the family photo at left with her mother, her brother, and her father.

The Frances had moved to this picture-perfect place soon after Diane was born on May 11, 1954. In 1957 her brother Mike came along. When both kids

were in school, Diane's mother, Dolores, worked in her husband's office, scheduling appointments, greeting patients, sterilizing instruments, sharpening needles—whatever was needed.

The office was just a few blocks from the France family home. The Frances lived on a hilly street with Diane's best friend, Joy Price, right next door and her other best friend, LuAnn Bybee, across the street. Diane was known around town as "the doctor's daughter." Joy's father worked at the county courthouse, so she was "the judge's daughter." LuAnn, a couple of years younger, was "the store owner's daughter." Her family owned the Sportsman Supply and the Walden Cafe.

Ten-year-old Diane France *(top left)* and her best friend, Joy Price, stand outside her father's medical practice with an office assistant, brother Mike, and Walden's only doctor—Dave France. The France house on Harrison Street *(above)* overlooked a grassy valley rimmed with mountains.

The kids at school jokingly called their little street Snob Hill, but the families who lived there weren't rich. In fact, Diane spent one summer weeding dandelions—two for a penny—to earn money for a bike.

~ *Mom's Rigid Rules*

Hard work was important in the France family, but setting a good example was paramount. As the doctor's daughter, Diane knew she was expected to behave well. The worst thing she could do was embarrass her family by earning bad grades or getting in trouble. It wasn't easy. In a small town, everyone knew your business. Parents watched over the other children along with their own.

Dolores France worked hard to keep her two children quiet and well behaved. Her rules were strict. Other kids shot marbles and played ball in the street, but Diane wasn't allowed to. For a time, her parents told her she couldn't even cross the street, let alone hang out there. If she wanted to go somewhere with Joy and LuAnn—over to the store or tubing on the river—the answer was often: "No—and that's final." Mom never gave a reason, no matter how many times Diane asked. The hardest rule was not playing with kids her mother didn't know or didn't approve of.

Ina Matney, the "minister's daughter" who lived directly across the street, next door to LuAnn, was close to Diane's age. Even better, Ina owned a pinto horse. Diane rode horses every chance she could and, though she was too shy to ask for a ride, Ina sometimes invited her over. Like Diane, the minister's daughter was a tomboy—only more so. Ina did some scary trick riding on that horse—hanging off the side, standing up, jumping on from the back. Diane learned a few tricks, too, but nothing that dangerous.

Then, one day, Diane's mother ruled that Ina's horse was off-limits—and so was Ina. While the girls fumed outside a closed door, Diane's mother was inside, telling Ina's mother that the two girls couldn't play together anymore. Diane wasn't even supposed to speak to Ina—which was pretty rude since she lived right across the street.

Diane (*left*) and Joy loved Bula (*the dog*), sledding, bike riding, swimming, and pretending to be spies with walkie-talkies. Joy said, "I felt like Diane was my sister more than a friend."

Diane felt deeply hurt. Then, she flashed with anger when, yet again, her mother wouldn't tell her the reason. Maybe Ina was too rough and reckless. Mom was always telling Diane to be more ladylike. Looking and acting like a lady was a duty, not an option, for the doctor's daughter.

Mike has it so much better, Diane thought. To her the message was clear: Girls cooked and cleaned and got dressed up while boys went fishing and worked in the yard. That was just how it was.

And that was just how it ought to be. Both her parents felt that way, but Diane soon discovered it was a mixed message.

~ *Science Stinks*

Joy and LuAnn played with Barbie dolls and loved to dress in groovy clothes. Joy wore fishnet stockings, go-go boots, a green miniskirt and vest, and a gold suede hat. LuAnn had a shag hairdo, teased to a poof in the back—just like Goldie Hawn on the TV show *Laugh-In*. Both Joy and LuAnn watched Elvis Presley movies to see what the girls in it were wearing.

They always look so great, Diane thought, but she just wasn't interested in "girlie things." She hated Barbie dolls so much that she buried them in little Barbie-sized haystacks. She put fresh grass clippings and a doll in a shoe box and then stuffed the box firmly with more clippings. When she turned it upside down, the block of grass popped right out like green Jell-O from a mold.

Diane loved science, especially anything to do with animals. Her kindergarten "boyfriend," Richard Smith, gave her a gopher that he found in the hayfields. She kept it until the day it popped up by surprise in the bed of grandmother Esther, who was visiting for a while. "No more gophers in the house," Mom ruled. Diane painted the gopher's tail blue so she could recognize it in the wild if she saw it. Then, she set it free.

She never got caught, but at the same time her intense interest in science—skeletons, animals, and chemicals—did not go unnoticed.

Wednesday afternoons and Saturdays—Dad's time off from the office—Diane would take his microscope into the hayfields. She never saw that blue-tailed gopher again, but she did come home with rocks, bones, seeds, antlers, and tadpoles. She put the tadpoles in a wading pool, and she and LuAnn watched the baby amphibians every day until they grew into frogs and hopped out.

Diane kept a chemistry set in the basement. Joy and LuAnn weren't interested in it until Diane introduced them to phenolphthalein [fee-nol-THAY-leen]. The colorless liquid turned pink when Diane added an antacid or some other alkaline. Then, it turned colorless again when she added a couple drops of something acidic—white vinegar or lemon juice, for instance. All three girls thought that was really cool.

Diane also entertained her pals by writing secret messages in "invisible ink"—ordinary lemon juice that becomes invisible when dry and reappears when held near a warm lightbulb. She felt especially proud one day when a stink bomb she made herself actually worked. She added wax to the recipe to make the bomb burn longer and steadier and then found ways to amuse herself with it. She never got caught, but at the same time her intense interest in science—skeletons, animals, and chemicals—did not go unnoticed.

One day her mother said, "Diane, why don't you be a nurse?"

"To heck with that!" said Dad. "Be a doctor. You don't want to be taking orders."

Diane didn't want to be a doctor—well, maybe an animal doctor—but her father's message sank in. Whether she did a "boy job" or a "girl job," she should reach for the top. Anything less was not good enough.

~ *Taking Aim at the Top*

At age 13 Diane learned how to shoot a BB gun and then aimed for first place at the 1967 Jaycees shooting contest. She won, beating out the kids of local ranchers and hunters. She also studied music, and Diane and Joy won a boxful of medals playing flute duets at music contests. In high school, Diane joined the drama club, the newspaper club, and the political science club. She worked an after-school job, stocking groceries for a dollar a day. On top of all that, she earned excellent grades, always striving to be the best at everything she tried.

All those achievements helped Diane win a town contest to travel to Finland for a couple of months in 1971, the summer before her senior year of high school. The host family had a girl Diane's age who spoke a little English, but few others did in the small factory town of Savio, about an hour north of Helsinki.

At first Diane was painfully homesick. She missed her boyfriend, Kip Varner, who was a year older and about to leave Walden for college. Kip was smart and good-looking and knew how to pull off a good practical joke. Diane admired anyone who could stir things up, and Kip's flair for mischief was a big attraction—even when the joke was at her expense. One night Kip and his best friend—Joy's boyfriend—somehow duped the girls into snipe hunting. Diane had never heard of a snipe but fell for it just the same. The girls pointed flashlights under quite a few bushes before realizing they were looking for imaginary beasts.

The 17-year-old Diane France strums her guitar at a gathering in Savio, Finland, where she lived with a host family during the summer of 1971. Savio is just north of Helsinki.

Diane smiled as she thought about how Kip eventually got a face full of payback. He and his two brothers trapped a live skunk, which promptly spun around and sprayed its three captors. The boys had to burn their clothes and take tomato juice baths to rid their bodies of the stench. *What a hoot!*

Diane missed her family, too, but not her brother's pranks. When she was 15 and Mike was 11, his voice hadn't changed yet and was high, like hers. One time Kip called and Mike answered the phone. Kip thought he was talking to Diane, his new girlfriend, and Mike went along with it. Mike even "made a date" with Kip—leaving Diane baffled when Kip showed up at the door. It didn't take Diane long to figure out what her little brother had done.

She also didn't miss her parents' strict rules, especially for dating. After a date, Kip made sure to bring her home on time, since Dad was usually watching at the window. If they stopped to talk in the driveway, Dad flashed the porch light on and off—his not-so-subtle way of telling Diane, *Get inside, now!*

~ *The Quest to Be Best*

Over the summer of 1971, Diane learned enough Finnish to talk to people. She discovered a taste for yogurt—a new food to most Americans then—and a passion for travel. Meeting new people and seeing new places was a grand adventure, she decided. Best of all, after returning to Walden at summer's end, she had great stories to tell. Now that she was under her parents' rigid roof again, her favorite story was about being ladylike.

Diane *(left)* pitches in on the family ranch in Finland. One of the host family's daughters later spent a summer in Colorado with Diane's family.

Diane told her folks that she had tried her best to be polite and well behaved. She worked hard at the host family's ranch, cutting and stacking hay by hand. In return the ranch hands taught her Finnish. They'd say a word, she'd repeat it, and the men would crack up. She figured they were laughing at her American accent, so she practiced the proper pronunciation. Later, on a bus tour with a church group, Diane tried out her new words. The church ladies looked horrified.

Diane was puzzled, but later she and her host sister figured it out. The "helpful" ranch hands had played a pretty good joke on Diane. Her new Finnish words were anything but ladylike.

In her final year of high school, Diane stepped up her pursuit of the top spot in the class. Grade-wise, she wasn't worried about 29 of her classmates. But she was neck and neck with the 30th, a boy named Brett. The competition made Diane work harder. In June, when the North Park High School class of 1972 was about to graduate, the final ranking of the 31 seniors was announced. Diane had come in second—just a few hundredths of a point behind Brett. Number two was an honor. It even came with a title: salutatorian. But all Diane could think was, *So close! How could I miss coming in first by such a tiny fraction?*

The salutatorian of North Park chose a college less than two hours from home. Colorado State University (CSU) in Fort Collins was a maze of multistory buildings crawling with thousands of students. The freshman class alone had far more people than the entire town of Walden.

In her North Park High School yearbook, the "talented" and "most dependable" Diane was pictured in bell-bottoms and a mini-skirt. She could also shoot a rifle, play the flute, and build stink bombs.

MOST TALENTED
Mike and Diane

NORTH PARK HIGH SCHOOL

COMMENCEMENT ACTIVITIES

ELEMENTARY GYMNASIUM

BACCALAUREATE MAY 28 - 8:00 P.

COMMENCEMENT JUNE 2 - 8:00 P

- 1972 -

MOST DEPENDABL
Brett and Diane

One advantage of a gigantic school was a wide choice of majors. Diane told her parents she was planning on a career in marine biology. She loved watching Jacques Cousteau movies about exploring the ocean, and she loved animals. To her, marine biology seemed like a logical choice.

Her dad pointed out, "Colorado has mountains, not oceans."

Diane didn't care. So what if she didn't know how to scuba dive. She would learn, just as she had learned to shoot and play the flute and speak Finnish. Besides, she could graduate from CSU in four years and move to California for an advanced degree.

At the age of 18, Diane France had played the role of the doctor's daughter as well as any teenager could. Now she was free to do exactly what she wanted. Maybe too free.

The year was 1972,

a time of **rebellion**

against **authority**.

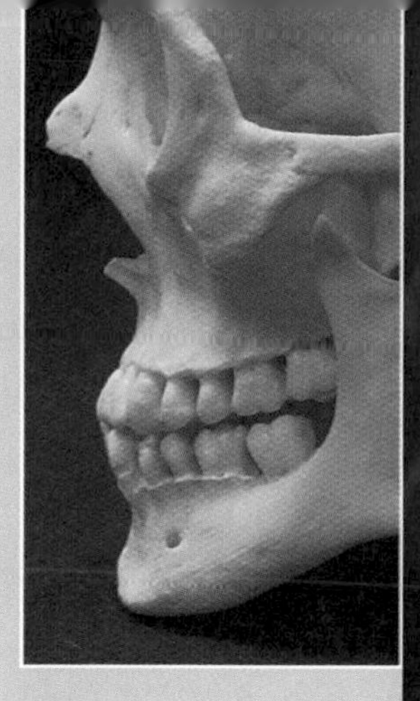

3

A Choice Education

So here she was: a small-town girl at a huge university, on her own for the first time, with an appetite for adventure. Diane France was giddy with freedom, but a little shell-shocked, too, at the craziness.

The year was 1972, a time of rebellion against authority. Bumper stickers warned, "Don't trust anyone over 30." Women fought discrimination, demanding "equal pay for equal work." Richard Nixon and all the president's men were sinking fast under the weight of the Watergate scandal. Students were staging demonstrations and marches against the war in Vietnam.

Between protests, it seemed to Diane, everyone partied all night and cut classes whenever they wanted. The general attitude was "If it feels good, do it." At CSU, Diane was no longer known as the doctor's daughter, so she thought, *Now I can do what everyone else does. I don't have to set an example*. She could study—or not. She could go to a party—or not. She could relax outdoors at the Horsetooth Reservoir as often as she liked. Whatever she did, it was her choice, and no one could say, "No—and that's final." She had a heavy load of science and math classes, but she figured she could slide by. She'd always been a good student. So mostly, Diane chose to have one heck of a good time.

University of California students *(opposite)* protest the Vietnam War (1964–75). The peace movement swept college campuses coast to coast, including CSU. Diane's interest in bones *(above)* developed in college.

At the end of the first term, the university kicked her out for poor grades.

~ A Second Chance?

Christmas break back in Walden was the most humiliating time of Diane's life. Her hometown had sent her off to college with the highest hopes, and now everyone in Walden knew she had failed. People treated her nicely, but she knew that they knew.

Diane's parents were astounded, angry, and embarrassed. Their daughter had graduated second in her high school class! They couldn't imagine what had happened.

Dolores wondered, "What is Diane going to do now? She'll have to look at other options."

She refused to accept failure as her only option. The university's ruling was final, and it was probably too late to undo the damage, but she had to try.

"Get kicked out of the family," Dave replied, only half-joking. "I don't know any other option."

Diane's parents refused to have anything to do with her for the entire break. To them the situation was clear. If you didn't graduate from college and get a good job, your life was a failure. Their attitude was extreme, but it didn't surprise Diane. Her parents had always expected her to excel. Until now she had tried her best to meet their lofty expectations. This time she hadn't even tried.

Diane was on her own, all right, but suddenly freedom did not feel so good. She had chosen not to study—a stupid mistake—but now she made a new choice. She refused to accept failure as her only option. The university's ruling was final, and it was probably too late to undo the damage, but she had to try. Immediately after the holiday break, she returned to campus, walked into the dean's office, and pleaded to be let back into CSU.

The dean had seen plenty of teenagers party their way out of college. He said, "Maybe you're just not ready for school."

"I *am* ready for school," Diane said firmly. "I really want to be here." She pointed out her stellar high school record. Suddenly, she was proud—and thankful—that she had graduated second. All of her achievements gave her firm ground to stand on in front of the dean.

"I can work hard again," she promised him. "I'll get good grades, if you give me another chance."

The dean allowed Diane back into CSU on probation, which meant she had to prove herself in the next term—or leave. The bad grades from the first term still counted, so she had acres of ground to make up.

~ Changing Course

Diane signed up for a full load of math and science again. She put on mental blinders to shut out the partiers and worked her fingers to the bone to earn A's and B's. She began dreaming again of life in California as a marine biologist. Then, in the last term of her freshman year, one class changed her life. That class was anthropology. She loved it so much that she signed up for an advanced class the next year. Her life's passion, she quickly discovered, was bones.

Piecing together a skeleton was like solving a jigsaw puzzle. From mouse to dinosaur, all vertebrates on this planet seem to be cut from the same mold. Their skeletons have the same basic bones in the same places, centered around a spinal column. The bones vary in size and detail, depending on how the animal moves and lives, but for the most part a femur is a femur.

Diane also became hooked on the game of identifying unknown bones. She and another student, Mike Gear, turned it into a friendly competition. The university's anthropology lab was lined with cabinets that stored unarticulated (unconnected) skeletons. These loose skeletons were filed by the year of collection—from archaeological digs, for example—but the individual bones were unidentified. Hour after hour, Mike and Diane would pick up a set of bones and quiz each other: *Is this human or nonhuman? Radius or ulna? Right or left scapula? Fourth rib or fifth?*

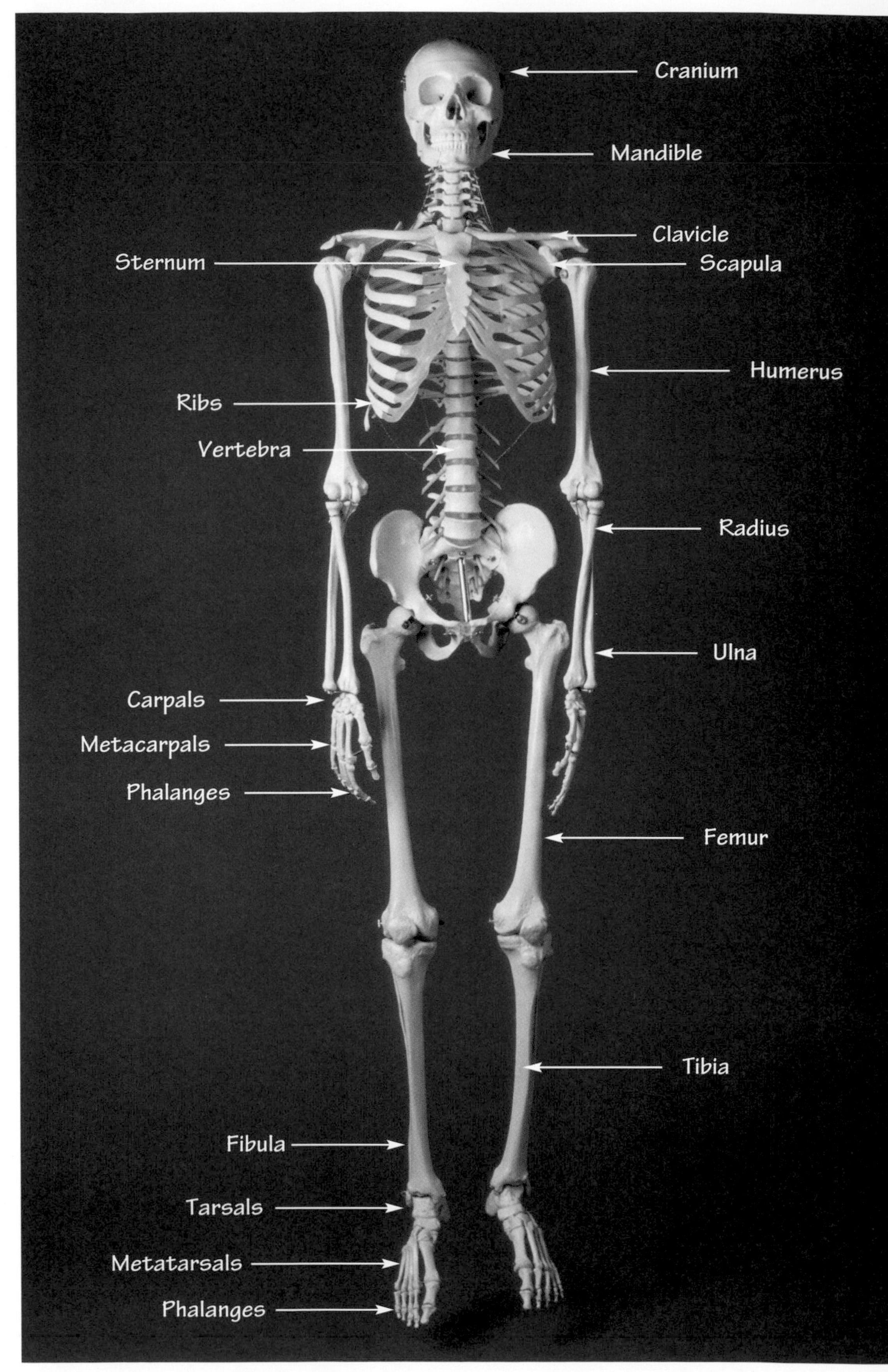

Cranium
Mandible
Clavicle
Sternum
Scapula
Humerus
Ribs
Vertebra
Radius
Ulna
Carpals
Metacarpals
Phalanges
Femur
Tibia
Fibula
Tarsals
Metatarsals
Phalanges

Beyond shape and size, Diane learned to look at texture. She saw that the head of a femur is smooth, which helps it rotate freely inside the hip socket. But the neck of the bone has striations—tiny parallel ridges—made by diagonal stresses. Weight bears down on this part of the bone at an angle. The shaft is hard and dense, strong enough to hold the bulk of the body's weight. Diane felt proud that she could now identify bits and pieces of bone by both texture and shape.

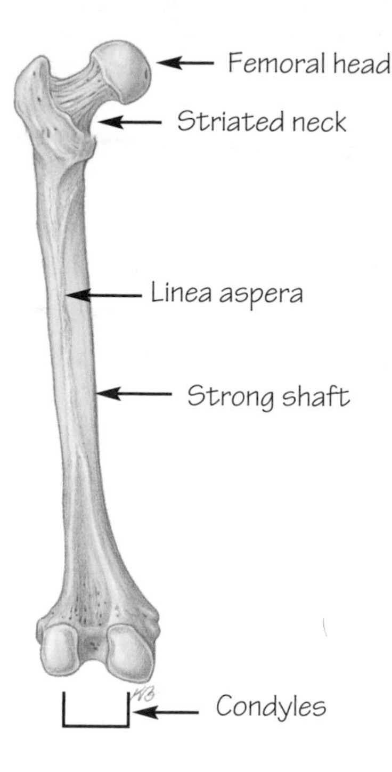

The femur is the biggest, longest bone in your body. The reason: Because you walk upright, it has to support your entire body weight. Feel the powerful muscles on the back of your thigh. They attach to the femur at a strip of rough bone called the linea aspera.

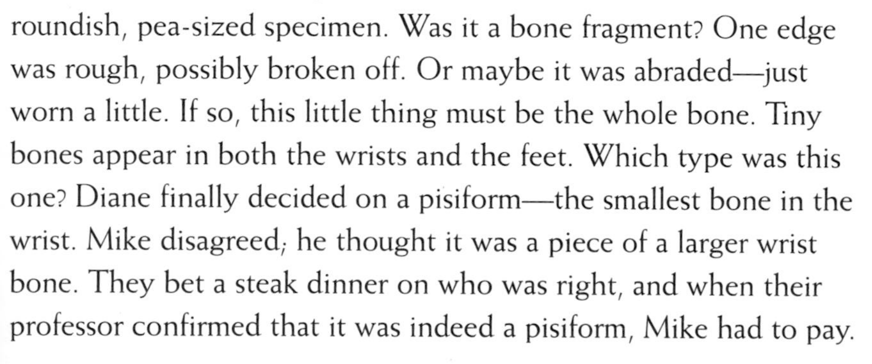

The ultimate challenge was a roundish, pea-sized specimen. Was it a bone fragment? One edge was rough, possibly broken off. Or maybe it was abraded—just worn a little. If so, this little thing must be the whole bone. Tiny bones appear in both the wrists and the feet. Which type was this one? Diane finally decided on a pisiform—the smallest bone in the wrist. Mike disagreed; he thought it was a piece of a larger wrist bone. They bet a steak dinner on who was right, and when their professor confirmed that it was indeed a pisiform, Mike had to pay.

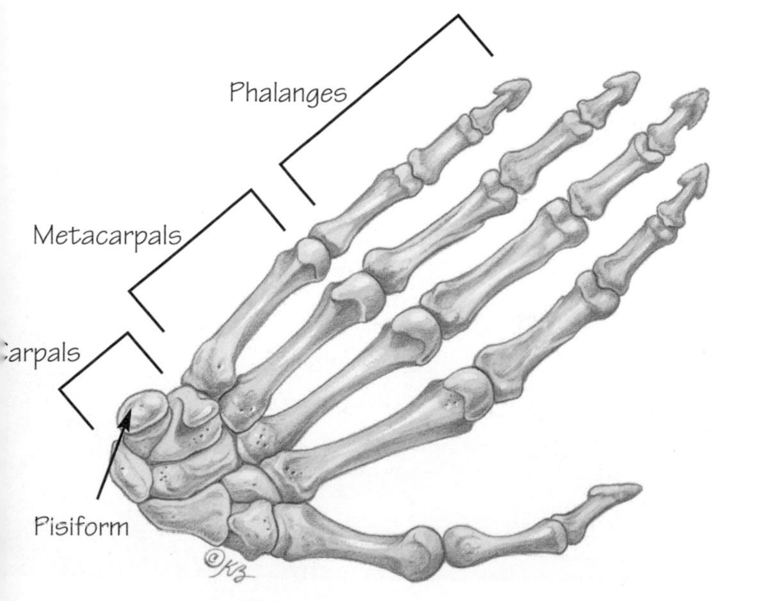

Look closely at the 27 bones in this hand. Now imagine them in a jumbled-up pile. Could you pick out the pea-shaped pisiform bone that forms a bump on your wrist? Could you tell one metacarpal from another? By playing a guessing game with a friend, Diane learned to identify all 206 bones in the human skeleton *(opposite)*. More than half of them are in the hands and feet.

~ *Life in a Skeleton*

Diane learned to study a skeleton and re-create a detailed picture of the person's life at the moment of death. She could say: "This person was a woman, most likely of European ancestry and in her early 20s. She stood 5′ 3″ to 5′ 7″ tall and probably bore at least one child. The bones showed no sign of disease or malnutrition, but a fracture of the left tibia was just starting to heal when she died."

All those facts don't pop out at a glance. Diane practiced hard to pick out tiny details and make very precise measurements of the bones. For example, she learned that having a child often causes scarring in the shape of two pea-sized pits, side by side, on the dorsal (inside) surface of the os pubis, a pelvic bone. A shin bone fracture is obvious if there's a callous. A callous is a bony bump that forms during healing but later disappears. If it is soft, the break happened shortly before death.

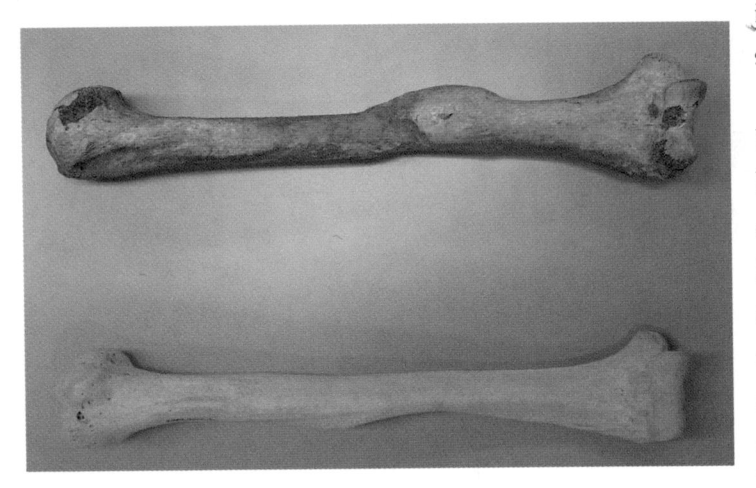

Which humerus bone broke and then healed? To find out, look for a callous—a bump like the one near the center of the top bone. It begins to form right away at the fracture site. If the callous is soft, the injury happened shortly before death. If it's hard, like this one, the bone had more time to heal.

Taller people generally have longer bones, but finding exact height is impossible because each body is different. Instead, Diane learned to estimate a range of heights. First, she measured the lengths of arm and leg bones in very precise spots. One way is to put the bone on an osteometric board—sort of like a foot measurer at a shoe store. Another way is to use calipers, which can look like either giant pinchers or

A femur fits snugly in an osteometric board, a tool for measuring bone length. The right wall slides back and forth to accommodate bones of different lengths.

How to Use a Bone to Estimate Height

Taller people tend to have longer leg bones. Yet people of the same height don't always have bones of the same length. Besides height, bones vary with a person's age, sex, ancestry, health, and other factors.

That's why forensic anthropologists can't use the length of a bone alone to determine a person's exact height. Instead, they use formulas to calculate a range of likely heights. The formulas are based on studies of thousands of skeletons of known height, sex, and ancestry. Here's how to use a femur to estimate the height of someone like Diane—a woman of European ancestry.

1. Measure the length of a bone in millimeters.

   Femur Length: 442 mm

2. Plug the number into a formula based on known heights and bone lengths for the same sex (female) and ancestry (European). Note that the result is in inches, not millimeters. (The conversion is built into the formula.)

   Formula: 0.11869 (442 mm) + 12.43 = 65" (5' 5")

3. Factor in the margin of error. For this formula, it's +/– 2.4", so there's an excellent chance the person was between 62.6" and 67.4" (5' 2" and 5' 6") tall.

a wrench with a sliding head. Diane then compared the bone lengths to skeletons of known height using math formulas. The formulas take into account differences in sex and ancestry (European, African, Asian, and so on), but, at best, they're accurate to within only a few inches. (*See box.*)

With all this bone detail swimming around in her brain, Diane couldn't help looking at living people in a whole new way. Her trained eye studied the shape of a face, the width of a nose, the distance between eyes. She was entranced by an unusually large brow ridge. She tried to imagine the skulls beneath these faces, their hidden features and shapes.

This cross section of a chicken bone shows what Diane learned from dinner table surgery. Beneath the bone's periosteum, a delicate membrane sheath, lies a hard outer layer of bone pierced here and there by foramina—tiny holes for blood vessels or nerves. Inside the bone, spongy tissue surrounds a core of dark red marrow.

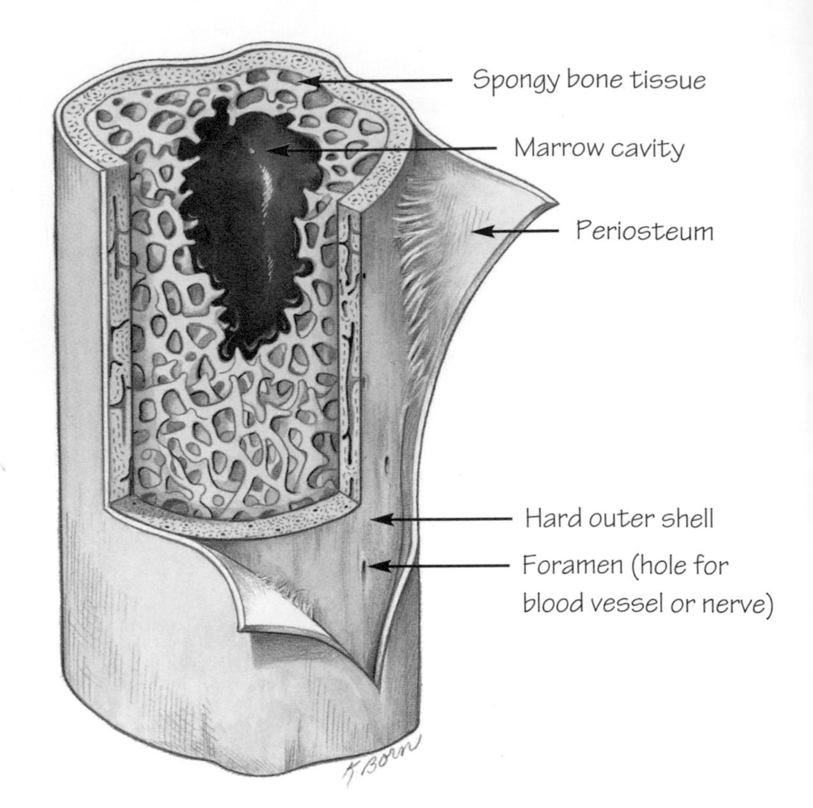

Diane's fascination with bones followed her to the dinner table, where a chicken dinner became a "fun with anatomy" lesson. Diane removed the fatty yellow skin and separated the meat-colored muscle layers with her fork. There, inside the cooked flesh, she found the rubberlike tendons and followed them to the spot on the bone where they were attached. She popped off the cap of white cartilage at the ends of a long bone.

Underneath the cap, she sometimes found an epiphysis [i-PI-fi-sis]. This tiny, disk-shaped bone was present only in young chickens—labeled "fryers" in the grocery store. By adulthood, the epiphysis was fused to the main bone.

Diane picked up a drumstick and took off the leg's fibula, a long skinny bone next to the tibia. Then, she scratched the tibia to lift the periosteum [pare-ee-OS-tee-um], a fragile membrane surrounding the bone. She snapped the bone in half to reveal the hard outer layers, spongy inside, and dark red marrow in the center. Diane admired the light, airy bone structure—an adaptation to flying.

Surveying the carnage on her dinner table, Diane thought, *Good thing I'm not at a restaurant!*

~ On the Money

Back in Walden for the summer of 1973, Diane told her parents her exciting news: She had decided to become a forensic anthropologist.

They looked worried and, Diane thought, very disappointed.

"How are you going to make a living?" her father asked. "You can't make money at that."

He was right. Forensic anthropologists weren't paid nearly as well as, say, doctors. The science of anthropology had been around for more than a century, but the forensic specialty—analyzing bones as evidence in an investigation—was brand new.

Diane didn't care about any of that. Forensic anthropology was in her heart and brain to stay.

To make money for college, Diane worked two summers at a sawmill, the highest-paying job she could find in her small mountain town. On her first day, Diane stood out like a sapling among mighty oaks. The lifers were all muscled men and women who worked with timber and buzz saws day after day. Many of them were missing fingers from on-the-job accidents. *There's a good reason this job pays so well,* she thought.

Diane's task was to manhandle lumber—lift it, carry it, stack it neatly. The first day was a killer. She had to carry pieces of wood that weighed almost as much as she did. A single 2 × 10 board (about 2 inches thick and 10 inches wide), 16 feet long, caused her small frame to shake, her legs to stagger, and her arm and back muscles to strain. Diane had to will herself to move each board. By the end of the day she was too sore to move. By the end of the summer, however, she was two shirt sizes bigger—and felt stronger and fitter than she ever had before.

~ Unfinished Sculptures

Diane's upper-body muscles—especially her biceps and triceps, her deltoids, and her latissimus dorsi—were bulging. Anyone could see these muscles, which most people called arm muscles,

Bones don't move on their own; muscles yank them around. Any kind of pulling force is called tension. Both tension and tendon come from a Latin word meaning "stretch."

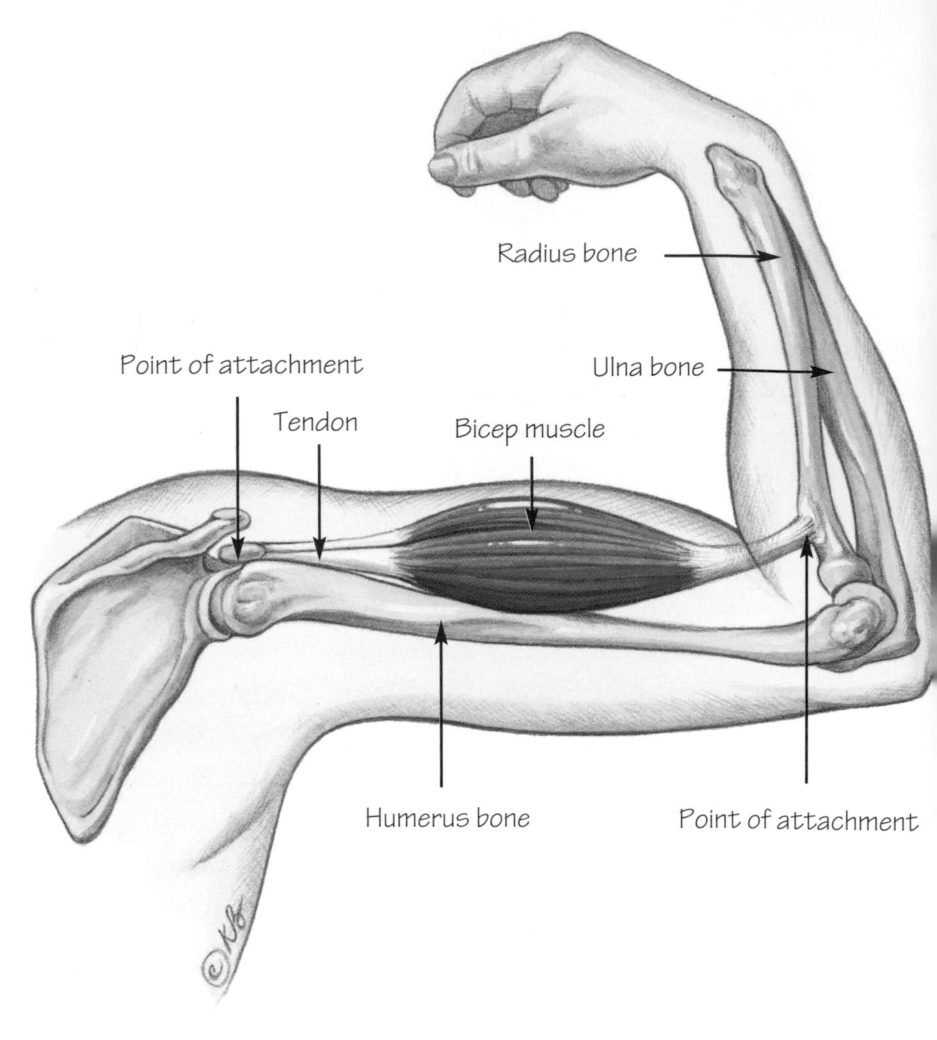

shoulder muscles, and back muscles. What no one could see, but Diane could picture in her mind, was how her bones had changed, too.

Her living bone had blood vessels, nerves, and busy bone cells worming through it. The bone tissue constantly remodeled and refined itself, like an unfinished sculpture. In response to stresses —pushes and pulls—or the lack of them, bone tissue thickened here, thinned out there. Pushes come from trauma, an impact injury such as the wham of a hammer. A hard blow can chip or splinter a bone; it can also damage the soft tissue around it. Bone reacts to trauma injuries by strengthening the tissue.

When it comes to pulling—a stretching force called tension — bones are as strong as steel. Muscles pull on bone every time they contract—tighten and shorten up. Diane flexed her arm like the Incredible Hulk and watched her bicep bulge. The flexed muscle was pulling on her arm bones at the points of attachment—

near the shoulder joint and just below the elbow joint. *(See diagram.)* The stress created small electric charges, which were triggering bone cells called osteoblasts to spring into action. Osteoblasts grab calcium from the blood and deposit it on the bone tissue at a point of stress.

After a summer of lifting lumber, this bone-building process—called the piezoelectric [pee-ay-zo-ee-LEK-trik] effect—had made Diane's bones denser and therefore stronger. But she knew that the biggest change was at those points of attachment. She pictured the muscle attachment points on the bone specimens back at the lab. They looked like rough spots on an otherwise smooth bone. Generally speaking, the larger the spot, the bigger the muscle had been. Diane now had some unusually big rough spots for a small-framed female. She'd never see them—barring serious injury—but it was cool to know they were there.

~ *Use It or Lose It*

When Diane returned to school and became a bookworm again, her muscles stopped overstressing her bones and the muscle tissue gradually shrank. This change signaled groups of cells called osteo*clasts* (rather than osteo*blasts*) to start tearing down excess bone tissue. The osteoclasts pulled calcium from the bone tissue and put it back in the blood.

No question, bones aren't just about biology, so Diane continued to take lots of physics and chemistry classes. For her forensic specialty—investigating crime scenes—she took a course in archaeology. Recovering artifacts at a dig site was surprisingly similar to recovering evidence at a crime scene.

At first she didn't like the archaeology teacher, Professor Cal Jennings. He embarrassed Diane in front of her classmates by calling her a "wallflower" who didn't speak up enough. Still, Professor Jennings did teach her how to set up a grid with stakes and ropes over a dig site. Using brushes and trowels, Diane learned to carefully excavate soil in a grid square, layer by layer. Using screens, she sifted out artifacts—tiny pieces of pottery, tools, and so on.

She carefully labeled each artifact by its location on the grid. She would later use this technique at outdoor crime scenes, where the goal was to sift out and recover bones, pieces of clothing, and any other object that might belong to either victim or killer.

~ *Graduation Days*

Despite her rocky start, Diane France graduated from CSU in 1976 with good grades and a bachelor's degree in anthropology. But she didn't stop there. She earned a master's degree in anthropology in 1979 and then went on to the University of Colorado at Boulder to pursue a Ph.D. in physical anthropology.

Dr. Alice Brues was Diane's favorite professor there. Alice had entered anthropology in the 1940s, when less than 5 percent of American women even went to college, let alone earned a science Ph.D. That took courage, confidence, and smarts. Here was a woman who had traveled the world

Diane beams upon earning her doctorate in 1983. With her is friend and mentor Dr. Alice Brues, one of the first women to earn a Ph.D. in anthropology (the inset photo shows Dr. Brues as a Bryn Mawr undergraduate). By the 1990s, two out of three Ph.D.s in anthropology went to women.

and could battle brains with anyone, anywhere. Yet she could also throw on a pair of jeans and hammer together a new deck for her home.

Most of all, Diane admired her professor's passion for science. Alice was so dedicated to comparing the shapes and sizes of bones that she carried a pair of calipers in her purse. When she met someone with an interesting head, she asked to measure it. A surprising number of people said *yes*.

Diane asked Alice how she had handled people who didn't think women should be scientists.

Alice replied: "If you don't recognize that view, it doesn't exist. If you let it bother you, then it's a problem."

Diane took that advice to heart. She focused on the men who encouraged, supported, and respected women. One of them, to her surprise, turned out to be Cal Jennings, the archaeology professor. A few years after graduating, she had gotten to know him as a person and a friend. In 1980, Diane decided to marry him.

College had prepared
Diane well for the
"goopyness" of her job.

Nothing, however,
had prepared her
for dealing with death.

4

MATTERS OF THE HEART

During a light snowfall in November 1980, Diane France married Cal Jennings at her family's cabin in a meadow rimmed with aspens and pine trees. *Yes,* she admitted to her college friends, he was the same Professor Jennings who had embarrassed her in archaeology class. True, she hadn't liked him at first. But a few years after her graduation, she and Cal had worked together on an archaeology survey in the Rocky Mountains. She began to see him in a much brighter light.

A mountain near the France family's cabin offered a beautiful getaway spot for Diane. The emotional impact of investigating tragedies like the gas explosion *(above)* led Diane to seek out quiet time to relax.

Cal had directed the field crew for the survey, including Diane and her dog, Sitka (Moki's predecessor). The crew was flown by helicopter into the mountains west of Denver each workday. They landed on a peak and hiked to the bottom, looking for any sign of past Native American presence. Then they were picked up and flown to the next peak. Sitka was mostly along for the ride, but she did one job better than all the humans: She was the first to hear the helicopter coming back and alerted her coworkers.

For fun Cal and Diane traveled as tourists to archaeological ruins. They also went on scuba-diving trips. Diane had recently learned how to dive, even though a career in marine biology was no longer in the picture.

Diane and Cal prepare to measure the precise distance to an object. Such measurements can help scientists map a crime scene.

Diane loved that Cal could talk intelligently about anything. They had an ongoing debate about whether chimpanzees have culture. Diane said, "Probably, if you look at culture as language, social behavior, game playing, tool use, and so on." Cal insisted, "Absolutely not. Do chimpanzees have politics, religion, and fine arts?" Though unresolved, the debate made Diane wonder.

Cal was 15 years older than Diane, and Diane's parents weren't crazy about that fact. But, gradually, they accepted him as part of the family. The newlyweds built a house in the country outside Fort Collins, and they both taught classes at CSU.

In 1983, Diane earned her Ph.D. after investigating the differences in humerus (upper-arm) bones between male and female skeletons. Her study focused on the points of muscle attachment—those same rough spots that had changed so dramatically during her "sawmill summer." After 11 years as a college student, she plunged into a career as a college professor and anthropologist at CSU. Archaeologists asked her to examine bones from their digs. Law enforcement agencies, coroners, and medical examiners sent the bodies of murder and accident victims to CSU's Human Identification Laboratory. Diane and other anthropologists examined the bodies for clues.

College had prepared Diane well for the "goopyness" of her job, as she sometimes called it. In anatomy class, both as a student and a teacher, she dissected a lot of cadavers. She also removed the soft tissue—skin, fat, nerves, blood vessels, muscles, tendons, ligaments—from corpses in order to examine clean bones. It wasn't her favorite thing to do. The process took several days of soaking the body in hot water. Still, Diane accepted it as part of the job.

Nothing, however, had prepared her for dealing with death. Her first mass fatality case was the hardest thing she had ever done in her life.

~ *Identifying (with) Individuals*

A week before Christmas in 1985, a gas explosion and fire killed 12 people at the Rocky Mountain Natural Gas Company in the small town of Glenwood Springs, Colorado. Diane arrived at the hospital just as a small team from the Colorado Body Identification team was packing up. This volunteer group of criminalists (forensic evidence experts) helped local coroners identify victims of fires, floods, plane crashes, and other disasters.

The team was able to identify seven of the 12 bodies from the gas company fire. Jack Swanburg, a veteran crime scene investigator, matched up fingerprints. Meanwhile, a forensic odontologist (tooth scientist) compared teeth to dental records. Unfortunately, five bodies were too badly burned for those techniques. These bodies became Diane's bodies—the ones she had to identify. Jack and his team said hello-goodbye and then left. They had to get back to Denver.

Not sure what to expect, Diane walked down into the cold basement morgue.

Diane used her skills as a forensic anthropologist to help identify the 12 victims of the Glenwood gas explosion. DNA testing, which identifies people based on their genetic material, did not come along until the 1990s.

Glenwood gas blast kills 10; two missing

GEORGE KOCHANIEC

An aerial photo of the site of a propane explosion Monday morning in Glenwood Springs shows firefighters battling the blaze as smoke pou

Fatal blast blamed on propane gas

Solving the Mysteries of a Mass Grave

Anthropologists sometimes face the challenge of identifying skeletons that are mixed together. These commingled skeletons could be from a pauper's cemetery, an ancient burial ground, a former war zone, an airplane crash, or a natural disaster. Whatever their origin, the goal is the same: Sort as many bones as possible into individual skeletons, then identify and analyze them. Any skeletal remains that can't be attributed to an individual are called common tissue.

A key first step is to figure out what is bone and what is not. Broken pottery and certain rocks can look like bone fragments. So can melted PVC (plastic) pipes that bubble up and reharden with a spongelike texture similar to bone tissue. One fast field test is to drop a little water on the object. Dry bone soaks up water faster than a thirsty sponge; most other materials don't.

Scientists also look at clusters—groups of bones bunched together. Chances are, they belong to the same skeleton, but not always. Common sense helps here: Two skulls or three scapulas (rather than two) means that at least two skeletons are present. A finer approach is to sort all the bones by type and by size: If two tibias are very different sizes

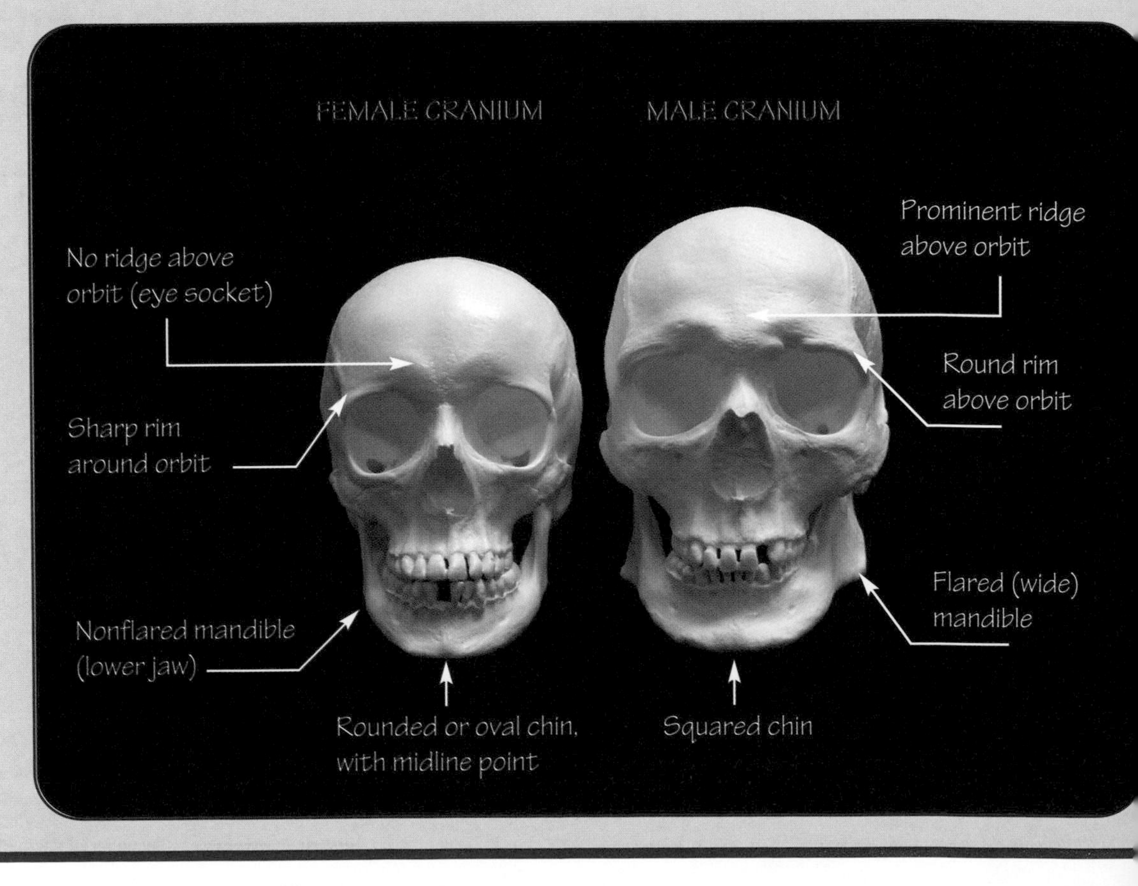

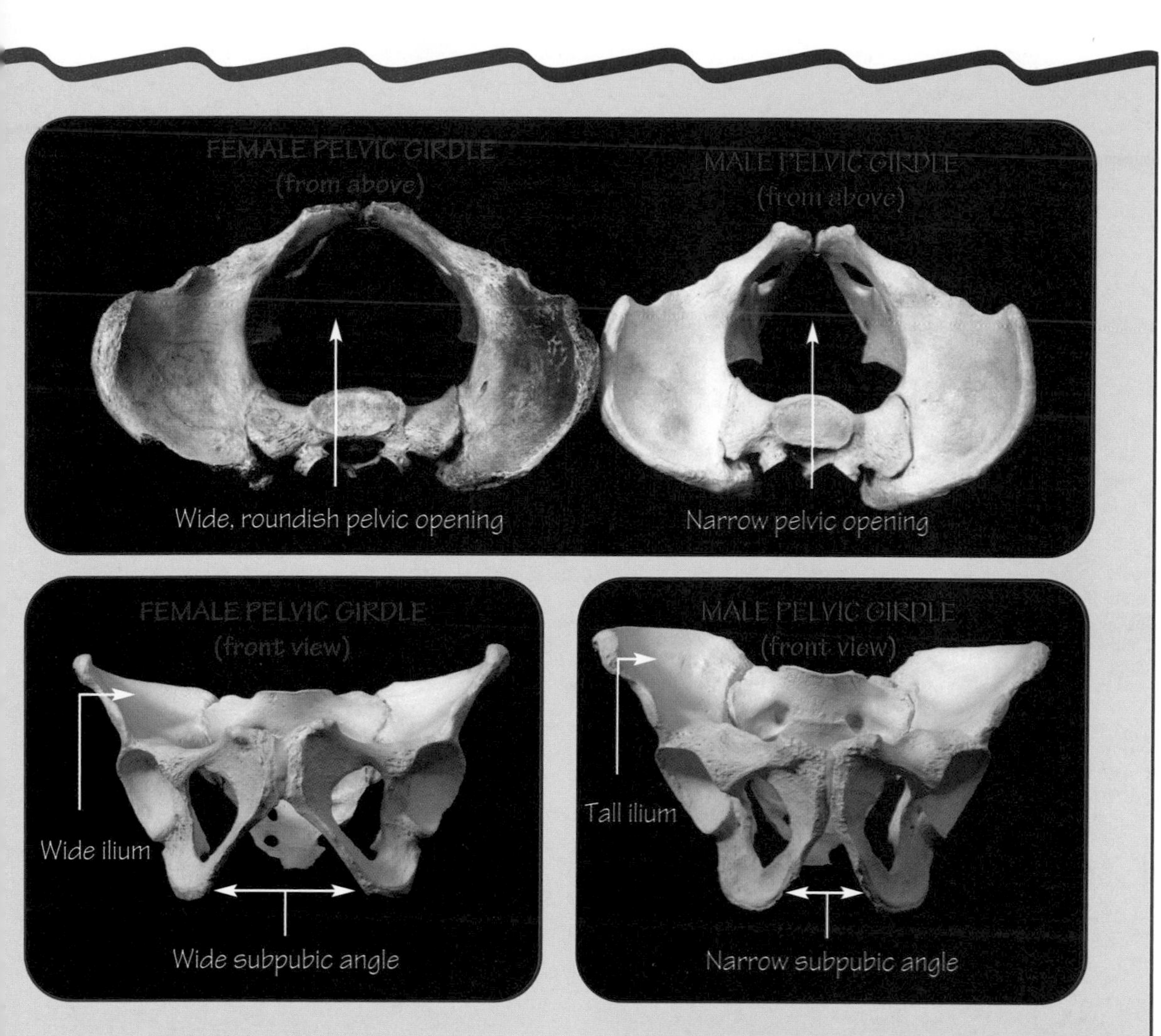

or they are both right-side bones, they belong to different skeletons.

Bone size is also a rough clue to a person's gender—men tend to be larger and more muscular (thus bigger boned) than women. But sometimes the reverse is true—as in a female basketball player and a male jockey—so there's room for error. The best way, by far, to tell males from females is by the pelvic girdle, or hip bones, because the pelvis of a woman has adapted to having babies. The second-best clues come from skull features.

In many cases, anthropologists can also estimate age, height, and ancestry (geographic origin). The more details they can determine, the more complete a picture they can create of who the individuals were and how they died.

One bone feature alone cannot determine the sex of a skeleton. A female skeleton could have skinny hips, for example, or a male skeleton could have gracile (fine and delicate) bones.

The idea is to consider all features as a whole, then decide whether the skeleton is more likely male or female. The pelvic girdle and the skull, in that order, usually show the sharpest differences between the sexes.

It was white and empty, except for some stainless steel tables. The burned bodies were stored in a separate walk-in cooler. She would have to unzip each body bag and examine its contents thoroughly by hand. The first instincts of most people would have been to run out and not look back. But Diane couldn't do that. It was a few days before Christmas, and the families of the victims were depending on her to identify their loved ones. At the moment, though, she identified more with the families—with their grief and pain. She also felt terrible for the rescuers, who, in such a small town, probably knew the victims.

This feeling of empathy was a double-edged sword. It motivated Diane to do her job because she desperately wanted to help these suffering people. Yet it also made her job emotionally hard to bear. Could she set aside these overwhelming feelings and get the job done?

Diane thought of a way. She conjured up a picture of a box in her mind. Inside that box she imagined Diane the Human Being—the sensitive person who felt deeply saddened by the deaths of the gas company workers. She closed the lid, wrapped the box with ribbon, and tucked it high on a shelf, out of her mind's eye. She would open that box later, she promised herself.

This feeling of empathy was a double-edged sword. It motivated Diane to do her job because she desperately wanted to help these suffering people. Yet it also made her job emotionally hard to bear.

She would learn how to deal with all those new emotions once the job was over. But for right now she was Dr. Diane France, forensic anthropologist—professional, well-trained, thorough—and she had a job to do.

Alone in the morgue, Dr. France worked late, then came back early the next day to continue the job. From time to time the coroner came down, asking when she would finish, telling her the families were waiting, pointing out that they wanted to bury their loved ones before Christmas.

Diane simply replied, "I'll get them done as soon as I get them done."

Few people had the knowledge and the training to do this job, and she felt both a duty and a desire to do it right. There was simply no room for error. She carefully separated bone from nonbone, then sorted and paired the bones by their size and shape. Once that was done, she could usually tell males from females and sometimes estimate age or other general traits. To help put a name to each body, she used the medical records of the missing individuals. If a radiograph (X-ray picture) showed a broken bone, for example, Diane looked for a healed fracture on the appropriate bone.

Diane passed the test. She identified the five unknown people. Altogether, the fire had claimed nine men and three women, ranging in age from 23 to 55, and now each one of the bodies had a name. The families, though grieving, were grateful to Diane.

~ *A Club of One*

Exhausted from her two-day ordeal, Diane headed home to Fort Collins. She watched the people shopping, eating, driving, talking, laughing. To her surprise she saw them differently now. They were living people, yet she kept seeing them as living bodies. She couldn't explain it to herself, nor to Cal. Only someone who had handled a mass fatality would understand, as if it were some kind of a special club. And at this moment, Diane was a club of one. She had worked the case alone and now she spent a few more days that way, taking a break from living people.

A couple of weeks passed, and the ribboned box stayed on the shelf. Diane told herself she just needed a little more time for the memories of the tragedy to fade. Meanwhile, she found plenty of ways to enjoy herself: shopping with her pals Joy or LuAnn, woodworking, gardening, playing with Sitka.

Three years later a piece of good news rose from the ashes of that fire. Diane was asked to join a group known as the Pig People. Despite the name, it was an honor.

*Jack Swanburg noticed right away that **Diane** was someone who didn't mind getting down and dirty.*

*"If there's something to be done," he marveled, "she gets on it and **does it**."*

5

The Pig People: A Grave Experiment

In 1988, Jack Swanburg, the fingerprint expert at the Glenwood Springs fire, went fishing for a forensic anthropologist. He needed one for a group of volunteer experts that he was helping to form. Unlike the Colorado Body Identification team, which assisted coroners, this new unnamed group worked at the request of law enforcement. Its mission was to help detectives find clandestine graves—any hidden place, on land or in water, where a murderer had secretly dumped a victim's body. If the group found a missing body, it would need a forensic anthropologist to help collect, identify, and examine the remains.

Diane emerges from her investigation of an abandoned Colorado mine shaft, a tempting place for a murderer to stuff a body *(opposite)*. A radio-controlled helicopter *(above)* can help investigators locate hidden graves.

One anthropologist who Jack contacted couldn't join but said, "There's a woman up at CSU who is new and who is good. Her name is Diane France."

The name meant nothing to Jack. But then he learned she had worked the Glenwood Springs fire a few years back. Jack pictured "the little lady"—young, petite, delicate looking—who had walked into the morgue just as his team was leaving. She had handled all that gore—bad smells, bad sights—by herself. He remembered feeling instant respect for her.

"We need this woman on our team," Jack said to his partners in crime fighting.

Over lunch Diane told Jack she was eager to join. She was now the director of CSU's Human Identification Lab, and she loved the idea of using science to help solve murder cases in the field as well as in the lab.

~ *Type A-Plus*

The group had about 15 scientists, criminalists, police detectives, and other specialists, including a dog handler and an aerial photographer. That number would soon double. Diane noticed that these varied volunteers had one thing in common: They all had type A personalities. They were strong, energetic, opinionated, and happy to take charge, like her. To succeed, though, they had to work as a team. So everyone dropped the titles—no "Doctor So-and-So" or "Captain Such-and-Such"—in favor of first names and a friendly atmosphere.

At the monthly meetings, one question was always on the table: What does a clandestine grave look like, and how do we find it? Everyone chimed in. Diane learned from the police detectives and criminalists how to think like a murderer. Killers are scared of getting caught, so they hide bodies in a variety of containers, wrappings, and remote places. However, the longer it takes to dispose of a body, the greater the chance of being spotted. So even in the wilderness, murderers tend to dump their victims near a road. Common sense says it's easier to drag a body downhill than up, to dig a grave in soft soil rather than hard, and to make the hole shallow instead of deep.

> They all had type A personalities. Like her, they were strong, energetic, opinionated, and happy to take charge. To succeed, though, they had to work as a team.

The geologists, botanists, naturalists, and other scientists tackled the question from a different angle: How does the presence of a dead body change the environment? What happens to the terrain, the soil, the vegetation, and the animal populations? For example, a decaying body immediately attracts decomposers—flies, ants, beetles, and other critters that feed on or lay eggs on animal flesh.

Scavengers—coyotes, bears, mountain lions—might also visit the gravesite, leaving behind footprints, hairs, and other clues.

Diane knew that as a human body breaks down it introduces fatty acids, gases, and other chemicals to the soil. Decomposition also releases heat energy. Geologists can bore a hole to get samples of soil and analyze its properties. Botanists can observe how soil chemistry and temperature affect plants growing over a gravesite.

There are also changes in soil density and magnetism. Untouched soil is packed more tightly than soil that has been dug up and crammed back into a hole. It also settles over time into a

Signs of a Hidden Grave

Nature changes a dead body over time, breaking it down and wearing it away. But how does a body change nature? The answer to that question is the key to spotting hidden graves. A decomposing body releases heat energy and chemicals into the soil, altering the plant growth above it. The odor of decay attracts certain insects that lay eggs on dead flesh—or feed on it. Scavengers such as coyotes sometimes leave footprints and other telltale signs of their presence.

To learn about these and other environmental changes, NecroSearch members study the gravesites of dead pigs *(inset)*.

north-south magnetic field; moving the dirt scrambles that field. Geophysicists can use various instruments to detect, measure, and map these soil properties without breaking ground.

The scientists all wondered what a clandestine gravesite looks like after weeks, months, and years. To investigate that question, the group had started an outdoor experiment in 1988, shortly before Diane came on board. The site was a field of prairie grass and scrub brush south of Denver. A botanist noted the plant growth. A geologist dug into the soil to examine the layers and types of rocks and minerals. A geophysicist measured the soil density, magnetism, and other properties. An aerial photographer snapped pictures of the field at dusk, when shadows made features more visible.

Volunteer experts Julie Kovats *(left)*, Diane France *(center)*, and Vickey Trammell carefully dig up evidence at a crime scene. The white grid allows Diane to map the location of each item found, square by square.

A radio-controlled helicopter *(left)* snaps aerial photos of a pig burial site. At dawn or dusk, long shadows bring out surface features, including the churned-up soil typical of a grave. The NecroSearch team looks for similar features when hunting for buried murder victims.

These and other measures became the "before" picture, baseline data of an area that had no graves. For an "after" picture—evidence of what happens after a body is buried—the group needed to bury some bodies. The use of human cadavers was restricted by Colorado law. So the scientists found the next best thing—pigs.

~ *Hog Heaven*

Pigs are similar to humans in size and body chemistry, especially because of their skin (furry animals decompose differently) and the ratio of fat to muscle. So the group buried dead pigs in ways that murderers sometimes bury their victims. A shallow grave imitated those "hurry up and don't get caught" burials. One pig had a "gunshot wound" (in this case inflicted after death). A "pig in a blanket" was wrapped up. A grave without a body was an experimental control—a way to compare which environmental changes were caused by dead pigs and which were not.

Insect expert Tom Adair set up a white bug trap *(below)* beside a dead pig, caged in to keep out scavengers. He studied which beetles and flies were attracted by the corpse. Their presence at a search site may point to a hidden grave.

The group gradually buried more dead pigs. Soon the experiment was dubbed Project PIG (for Pigs in Ground), and the members became known as the Pig People. They later elected to adopt a more dignified name: NecroSearch International. It was a clever choice: *Necro* comes from the Greek word for "dead body."

To Diane France, Project PIG was scientifically helpful. The data from the site helped the group identify many of the environmental signs of a hidden grave. But "hog heaven" was putting the information to use in real murder cases. Each case started with a request from law enforcement for NecroSearch to look for a missing body. Diane saw every outing as an adventure—and a chance to catch a cold-blooded killer.

Jack Swanburg noticed right away that Diane was someone who didn't mind getting down and dirty.

"If there's something to be done," he marveled, "she gets on it and does it."

A case in point was the 1993 search for a missing man in an oil field near the town of Kimball in western Nebraska. Jack and other NecroSearch members took turns holding Diane's ankles while she dangled head first into a "rat hole." The hole had been made by oil drillers, not rats. It was about 18 inches in diameter, Jack estimated, and once held giant drill bits. The hole had been filled with dirt—and perhaps the body of a murder victim. The NecroSearch team now had to dig it out.

"What's a little thing like you doing in a place like this?" "The same thing as you," Diane replied automatically and then went back to work.

Diane, at barely 5′ 4″ and 110 pounds, was the smallest person on the team, so she had volunteered to search the hole for "Shorty." Shorty was an achondroplastic dwarf reportedly killed in a dispute over money. Achondroplastic means that, as a child, his cartilage had not converted into bone, causing dwarfism, or stunted growth. His skeleton, if Diane could find it, would be easy to identify. Hanging upside down and using a small shovel, she carefully scooped the dirt out of the rat hole.

When she came up for a break, a burly Nebraska state trooper shook his head and asked, "What's a little thing like you doing in a place like this?"

"The same thing as you," Diane replied automatically and then went back to work.

She'd heard that question many times before—and usually in those exact words. Taking the advice of her mentor Alice Brues, she just ignored it. Two hours later Diane reached the bottom of the rat hole. No Shorty. The team even searched a mouse hole—a smaller drilling hole. Still no Shorty.

Buried in her work, Diane scoops out handfuls of dirt in search of "Shorty," a murder victim. A backhoe could do the job faster and easier, but it would destroy prized evidence.

The NecroSearch team eventually had to give up and return home, disappointed. The only consolation was that the search had ruled out a possible burial site. Shorty never did turn up, and without a body, the murder went unsolved.

Finding one hidden corpse in a vast landscape turned out to be harder than finding a needle in a haystack. In fact, during the first few years of cases, NecroSearch had zero hits and all misses. That explained why the team's first success was followed by war whoops and a human stampede.

During the first few years of cases,

NecroSearch had zero hits

and **all misses**.

That explained why the team's

first **success**

was followed by war whoops and

a human stampede.

6

Killer Secrets Unearthed

One August day in 1992, Diane was taking baby steps, eyes to the ground, on Kebler Pass, a steep, wet roadside slope in Colorado. Doing the same nearby were other NecroSearch members and a group of Diane's anthropology students from CSU. Diane held onto branches with one hand to avoid sliding downhill. Her other hand alternated between slapping mosquitoes and probing the ground for bone fragments, pieces of clothing, or any unnatural object.

The search centered on a spot where a hiker had found a long, dark-brown braid in 1979. Michele Wallace, a 25-year-old photographer with braided hair, had disappeared in the area in August of 1974, five years earlier. She had spent a few days camping and hiking with her dog, Okie, and taking pictures. Then, as she headed back to town, she gave two stranded men a lift in her bright red station wagon. Michele was last seen at a bar in Gunnison, Colorado. Her German shepherd turned up a few days later, roaming free on a ranch. Her red station wagon, camping gear, and camera ended up in the hands of one of the hitchhikers, a convicted thief named Roy Melanson. Roy told police that Michele had given him the goods. He claimed he knew nothing about her disappearance.

Diane measures the location of a piece of evidence in a grid square *(opposite)* at the Michele Wallace crime scene. The depth of the item below ground is just as important as its distance from the edges of the square. In the photo above, Diane works another crime scene with NecroSearch.

Michele Wallace loved hiking with her dog, Okie. After Michele disappeared in Colorado in August 1974, Okie turned up, alone, at a local ranch. Michele's camera, with this photo on a roll of film inside it, landed in the hands of Roy Melanson, a convicted thief.

Now, almost exactly 18 years later, Detective Kathy Young of the Gunnison County Sheriff's Department, was on the case. Like the original investigators, she suspected that Roy was the killer but didn't have enough evidence to convict him. Without a body, a murder case is very hard to prove. After hearing about the Pig People, she asked NecroSearch to help. Vickey Trammell, a NecroSearch botanist, examined the braid. In it she found needles of subalpine firs, trees that are adapted to living at very high elevations. These pine trees also prefer cool, damp, north-facing slopes that haven't been scarred by forest fires in recent years—another clue about where to search. The braid had little dirt and the surface was sun bleached, so Vickey believed the body had not been buried. She and Cecilia Travis, a NecroSearch naturalist and geologist, further reasoned that scavengers had probably scattered the exposed remains.

On the first day atop the high, north-facing mountain slope, Diane and the others found nothing. The next day they returned to search some more. Around noon they climbed back up the road for a break. Minutes later a faint voice rang from the woods below.

"Yoo-hoo! I found it!" Diane recognized the voice as Cecilia's but couldn't see her through all the pine trees.

"Found what?" someone yelled back.

"Michele!" Cecilia said, a little louder.

~ A Two-Legged Stampede

Yelling and whooping, the team stampeded toward Cecilia's voice. They ran and fell some 400 feet down the slippery slope to where she was standing. Cecilia pointed to what looked like a giant white mushroom growing in the middle of an animal trail.

Like Cecilia, Diane saw at second glance that this was no mushroom. It was the top of a cranium, the round upper part of the skull that houses the brain. A streak of sunlight, shining through the dense leaves, landed squarely on the row of upper teeth. A gold tooth sparkled. Diane remembered from dental records that Michele Wallace had a gold tooth. From the skull's shape and other characteristics, she noted, it was female, likely of European ancestry, and an adult—definitely not a child or teenager. The chances were very good that these bones belonged to Michele.

Diane points out the missing mandible (lower jaw) on the skull of Michele Wallace, whose remains were found by the NecroSearch team in August 1992. Gravity had caused the cranium to roll down a steep slope. The mandible was found higher up the hill.

Everyone—the NecroSearch team, the detective, and the students—cheered and high-fived each other. Finally! NecroSearch had its first success. The team found a clandestine grave hidden for 18 long years. What a coup!

Diane stopped celebrating and frowned. She looked at all the people swarming around. She realized that, like two-legged bulls, they had just trampled a crime scene. *Who knows what evidence we destroyed?* she worried. Michele Wallace deserved better.

Diane shouted for everyone to freeze. She dropped to her hands and knees and immediately began creating a walk line, a narrow path for walking to and from the crime scene without treading on clues. Diane carefully clipped away the foliage while searching the path for any evidence. Meanwhile, Kathy, the detective, alerted the coroner that a body had been found. Sheriff's deputies roped off the area and firmly shooed along curious drivers on the road.

Diane France and Steve Ireland, a NecroSearch archaeologist, hunt for evidence at the Michele Wallace crime scene. It is vital to collect and catalog every recovered clue—even a single strand of hair.

The next day Diane France and Steve Ireland, a NecroSearch archaeologist, organized six-person search lines to comb the area for more evidence. Each team roped itself together. The rope kept them from rolling downhill, possibly destroying evidence. The searchers lined up sideways to the road, with one end of the line at the top of the slope and the other end near the bottom. They dropped down on hands and knees and crawled, slowly and all together, across the damp slope, peering at every inch of ground before them.

Anyone who touched the items they found would become a link in the chain of evidence—a minute-by-minute record of who had custody of the materials. The fewer links in the chain, the simpler it was to present the evidence at trial and the less risk there was of losing something. So the search teams flagged items for detectives and criminalists to collect. Over several days the volunteers found foot bones, two thigh bones, a pelvis, six ribs, a vertebra, part of the sternum, and other bone fragments.

Meanwhile, Diane and Steve set up an archaeological grid with stakes and ropes. The idea was to pick through the soil for buried evidence. The tiniest item—even a thread—could turn out to be the "smoking gun" that cracked the case. With one person assigned per one-meter grid square, Diane and the other diggers used brushes, trowels, and wooden bamboo sticks to carefully remove 10 centimeters of soil at a time. (Archaeologists use only the metric system.) The bamboo sticks helped clear away dirt around a bone without cutting or scratching it. Accidental marks could be mistaken for injuries to the body. Cecilia Travis and other volunteers sifted the soil through fine screens to catch anything the diggers might have missed. The screens yielded a button, a zipper, a clothing clasp, a tube of lip balm, a yellow thread that possibly came from denim jeans, and a hair.

One big reason she loved her job was that she could speak for the victim. These bones were like Michele's last words. They told the story of her life—and of her death.

~ *The Victim's Voice*

A few days after Michele's remains were discovered, Diane examined the bone evidence for clues. One big reason she loved her job was that she could speak for the victim. These bones were like Michele's last words. They told the story of her life—and of her death. As a forensic anthropologist, Diane had been trained in how to listen to these voices from the grave.

A forensic odontologist had already matched the teeth with Michele's dental records, so there was zero doubt about the identity of the bones. Instead, Diane focused her exam on the crime itself. She wanted to know how this young woman died and what happened to her body afterward.

Unfortunately, Diane could find no evidence of what killed Michele. There were no bullet holes, knife wounds, or other visible injuries. It was still possible that Michele had been shot or stabbed or strangled, but the skeletal remains weren't complete. Time had erased too many clues. The cause of death was, therefore, "inconclusive."

On the other hand, it was clear how and where the body had been dumped. The vertebrae and ribs were found at the base of a tree about 25 feet below the road. Diane concluded that the body was tossed from the roadside, rolled downhill on its own, and came to rest at the tree. Scavengers then probably scattered some of the remains.

Over time, gravity must have caused the skull to roll farther downhill to the spot where Cecilia found it. This fact came to light accidentally at the crime scene. Twice, someone knocked over an empty bucket by the road; both times it tumbled downhill, stopping close to the spot where the cranium has been found. The "bucket test" became part of NecroSearch's growing list of clues for finding a missing body.

In 1994, Diane testified at the murder trial of Roy Melanson. It was her first time on the witness stand. She was nervous, but stuck firmly to the scientific facts. She didn't speculate about who killed Michele or why. She was there only to explain what the bones told her.

The Michele Wallace search team included scientists, detectives, criminalists, and volunteers from Diane's anthropology class. Diane appears at lower right, holding a shovel. The two young boys, sons of Detective Kathy Young *(second from right)*, made sure Moki, the malamute, did not wander into the crime scene.

The detective, the deputies, the criminalists, the NecroSearch team, the volunteer searchers—everyone's hard work paid off. Nineteen years after the crime, Roy Melanson, now in his mid-50s, was convicted of first-degree murder and sentenced to life in prison.

That first success put NecroSearch in the news. Dozens of law enforcement requests for the Pig People came streaming in. The volunteers could not take every case, but the ones they did take added to the NecroSearch store of knowledge. Making mistakes, as they would soon find, was especially informative.

~ *The Nose Knows*

In the mid-1990s, Diane and Steve Ireland, the NecroSearch archaeologist, headed a team looking for the body of Cher Elder. The 20-year-old waitress had disappeared near the tiny mining and tourist town of Empire, Colorado, in March 1993. Once again, the detective, Scott Richardson, had a murder suspect but no body. Cher had last been seen at a casino with Tom Luther, a convicted violent criminal and a suspect in two other murders.

Along with the detective, three bloodhounds named Amy, Yogi, and Becky worked the case at various times. They are "decomp" dogs, trained to detect the scent of decomposition. Amy and Becky are handled by Al Nelson, a NecroSearch member. Amy started in Empire and led Al to a waste treatment pond. The sludge was drained and the mud dug up, but no body was found. Becky pointed investigators to an abandoned mine. It also proved to be a dead end. Then, she paused at a large pile of rocks—one of many on the slopes off Highway 40. The pile was quite a distance uphill from the road—an unlikely place for a killer to drag a body. But just in case, Steve lugged away a few rocks and drilled a test hole in the frozen soil. He saw no signs that the soil had been disturbed.

Cher Elder, age 20, was last seen in March of 1993 in the company of a violent convict named Tom Luther. In prison, he had told a cellmate that his next victim would never be found. Diane's team proved him wrong.

In February 1995, a friend of Tom Luther spilled the beans in exchange for a lighter sentence on an unrelated crime. The friend mentioned a rock pile off Highway 40. Could it be the same rock pile that the NecroSearch team had stood on, drilled into, and abandoned? Diane, Steve, and other NecroSearch members took a look. This time the team included Diane's husband Cal, who had recently joined NecroSearch as an archaeologist. The team drilled deeper into the pile than before, even though killers normally don't dig far into hard soil. Diane laid on her stomach and stuck her nose in the hole. Right away she smelled decomposition, most likely human. The all-too-familiar odor was pungent and earthy.

The team excavated the rock pile layer by layer until, sure enough, they reached the body of Cher Elder. Diane noticed that the experience made a lasting mark on Cal. She saw it in his eyes. Her husband was used to digging archaeological ruins with dry, fleshless bones that were hundreds or thousands of years old. Cher Elder had been a lively young woman just two years before.

The snowy rock pile at top looked like any other mound of stones left behind by Rocky Mountain miners, yet it proved to conceal Cher Elder's grave *(above)*. Her killer, Tom Luther, had carried and stacked each large rock, one by one, to hide the body.

At the next NecroSearch meeting, the members reviewed their mistakes.

One, bloodhounds can't detect odors well through frozen ground.

Two, some killers don't follow commonsense rules, so expect the unexpected.

And three, look at the lichen. This colorful fungus, the team noticed after the fact, covered the tops of some rocks and the sides or bottoms of others. Obviously, the killer had picked up the rocks and dropped them every which way as he buried the body. On other nearby rock piles, the lichen pattern was uniform.

Despite these lapses, NecroSearch had found the body. In 1996, Tom Luther was convicted of murder and sentenced to 48 years in prison.

The cases kept coming. In less than 10 years, NecroSearch looked for dozens of clandestine graves—some hits, some misses. Then, one November day in 1997, a sea captain appeared before the group proposing to whisk Diane and her friends to a faraway place and a long-ago time. Once upon that time, the bodies of a beautiful grand duchess and her little brother, a crown prince, disappeared. Could NecroSearch please find them?

Diane couldn't believe her ears. *Was this fairy tale for real?*

Once upon that time,
the bodies of a beautiful grand duchess
and her little brother,
a crown prince,
disappeared.

Could NecroSearch please find them?

7

The Royal Bones

Taking the sea captain's case broke all the rules, and Diane knew it. NecroSearch took cases only on behalf of law enforcement. The captain, Peter Sarandinaki, was not a detective, and he wasn't trying to solve a murder. In fact, the two missing children had been dead for nearly 80 years. True, Peter explained to the group, but these children were *Romanovs*! He said the name with such reverence he could have been talking about saints. But Diane was pretty sure that the Romanovs were merely royals.

Czar Nicholas II, his wife Alexandra, and their five children sit for a royal family portrait *(opposite)*. From left to right, the daughters are the grand duchesses Marie, Tatiana, Olga, and Anastasia. Seated in front is Alexis. Above are skeletal remains of the family.

The Romanov family reigned over Russia, the world's largest country, for 305 years. When communists came to power in 1917, Nicholas was forced to give up the throne—and any claim to it by his only son, Alexis. Then, on the night of July 17, 1918, the last ruling Romanovs suffered a horrible fate. Communist soldiers—known as Reds—herded Czar Nicholas II, his wife Czarina Alexandra, their five children, and four aides into a stone cellar. The soldiers executed all of them. Peter said that his great-grandfather and grandfather, both White Russian officers, had arrived seven days too late to save the royal family.

Diane nodded. *That explained Peter's reverence.*

In 1991, Peter added, nine of the bodies had been unearthed in a mass grave near the city of Ekaterinburg, in Siberia. Diane did the math. Sure enough, two bodies were missing.

"A daughter and the only son, Alexis, have never been found," Peter said. He hoped NecroSearch could reunite all 11 skeletons of the royal family before their funeral set for July 1998, less than a year away.

Romanovs, Diane thought. She knew what most Americans knew about them—that they were royal Russian legends. Movies had been made and history books had been written about them. Who could resist joining this exciting quest?

Clark Davenport and John Lindemann, for two. These NecroSearch scientists took a less romantic view. They argued that the Soviet Union—a superpower collection of communist states headed by mother Russia—had collapsed just six years before, in 1991. The new countries that emerged, including Russia, were cash strapped, unstable, and dangerous.

"Besides," Clark added, "NecroSearch isn't about making history. It's about solving crimes."

Diane agreed. There were fresh graves to find, grieving families to help right here in the present. But she couldn't resist—this case was too thrilling to pass up. Here was a chance to be part of history, and she just *had* to take it.

~ *Night and Day*

In February 1998, Diane, Peter, NecroSearch geophysicist Jim Reed, and an anthropologist named Tony Falsetti flew to Russia. They arrived at 2 A.M. in Ekaterinburg, a city in the Ural Mountains of Siberia, the coldest region of Russia. The Americans knew they had entered a strange new world the moment they picked up their luggage. All of the address tags were gone. Every single one.

"Maybe someone stole them as souvenirs," Jim wondered. "Or to sell as trinkets in those sidewalk stalls."

A familiar sight in Siberia: the main street of Ekaterinburg buried in snow.

"Maybe," Diane said, but she was too tired to care. She couldn't wait to collapse into a nice warm bed.

Unfortunately, as a badly shaken Russia rebuilt itself in the 1990s, tourism wasn't at the top of the to-do list. Check-in at the "luxury" hotel—the best in town, Diane was told—took several hours. Worse, the hotel had no heat, so the foursome shivered in the lobby bar while waiting for their rooms.

"Traveling to Siberia in winter might not have been the smartest timing," Jim said.

Diane smiled. "No foolin'!"

Diane's room turned out to be as cold as the lobby—and had no hot water besides. She piled on clothing and tried to grab an hour of sleep. The wake-up knock came while night was still turning into day.

~ A Cold Shoulder

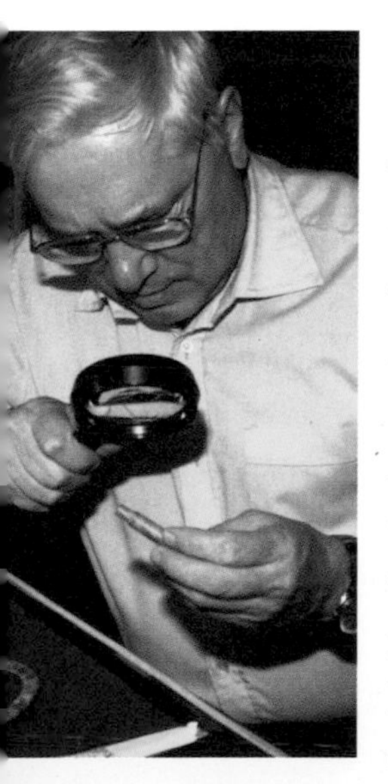

Russian geologist Alexander Avdonin discovered the gravesite of the Romanov royal family in 1979. Fearing the Soviets would destroy it—and himself—he kept its location secret for 10 years.

The mission of the American team's trip was to scope out possible gravesites and come up with a search plan. Then in the spring, a NecroSearch team would return to look for the missing bodies. Peter's contact was Alexander Avdonin, a silver-haired geologist in his 60s. Avdonin, as everyone called him, wore thick glasses balanced on a large ruddy nose. He had found the nine bodies—five Romanovs and four aides—in 1979 in a meadow near the city. It was a logical starting place to look for the two missing children.

Normally a 30-minute drive, the trip stretched into two agonizing hours. The car inched through heavy traffic and then had to detour around a snowy road. Making matters worse, there was another type of chill in the air. Avdonin was friendly to Peter, who spoke Russian, but Diane sensed a cold shoulder toward the rest of them. Jim and Tony felt it, too.

"Shaking hands with that man is like holding a dead fish," Jim said.

Diane thought Avdonin was especially rude to women. Some men still didn't take female scientists seriously. Diane couldn't believe how Avdonin barked at Galina, his wife. She was obviously translating his gruff Russian into candy-coated English. Her soft words were a sharp contrast to Avdonin's aggressive tone.

Tony figured Avdonin was wary of Americans trespassing on Russian turf. In 1992, the Russian government had invited another American team, led by forensic anthropologist Bill Maples, to help identify the nine skeletons found in the gravesite. The Americans openly refuted several findings of the Russian scientists, who weren't happy about being told they were wrong. After Bill Maples died, Tony had inherited the case files that contained the disputed evidence.

~ *The Royal Gravesite*

In 1989, Avdonin revealed the Romanov burial site in a meadow outside Ekaterinburg *(shown right)*. But it wasn't until the breakup of the Soviet Union in 1991 that the government finally allowed scientists to dig up and examine the remains. They found only nine skeletons; two Romanov children remain missing.

At last the car pulled off the road and stopped. Diane stepped out and felt a shock of cold air. The snow was so deep—four feet, she estimated—that it nearly buried a large Russian cross marking the gravesite.

A hired snowplower failed to show up, but Jim saw a bigger challenge: A large steel pipe cut through the meadow, and an

electric railroad with high-voltage wires ran beside it. Both gave off electromagnetic emissions that would muck up his magnetometer and other geophysical instruments. As Jim pondered how to overcome that handicap, he pulled out his digital camera to snap a panorama of pictures. It was so cold that the batteries didn't work. He warmed them in his armpit and popped them back in the camera. After a few shots, they quit again. To take the full panorama, Jim had to shuttle two sets of batteries back and forth between armpit and camera.

Jim explained to Avdonin that he planned to use the photos to generate a computer map of the terrain. Later, he would enter into the computer data collected about the area's botany, entomology (insects), soil properties, and so on. The computer would then compile and compare the data and indicate on the map the most probable spots for a hidden grave.

For decades, geologists had been using computer mapping programs to locate oil and minerals, but Jim's software was specifically designed for finding clandestine graves. Avdonin was impressed.

The next stop, he announced, was the morgue.

~ Battered Bones

In the morgue, Avdonin led the group to a door with an armed guard. The geologist removed a wax seal—a Russian-style security lock. Then, he beckoned the four Americans to step inside.

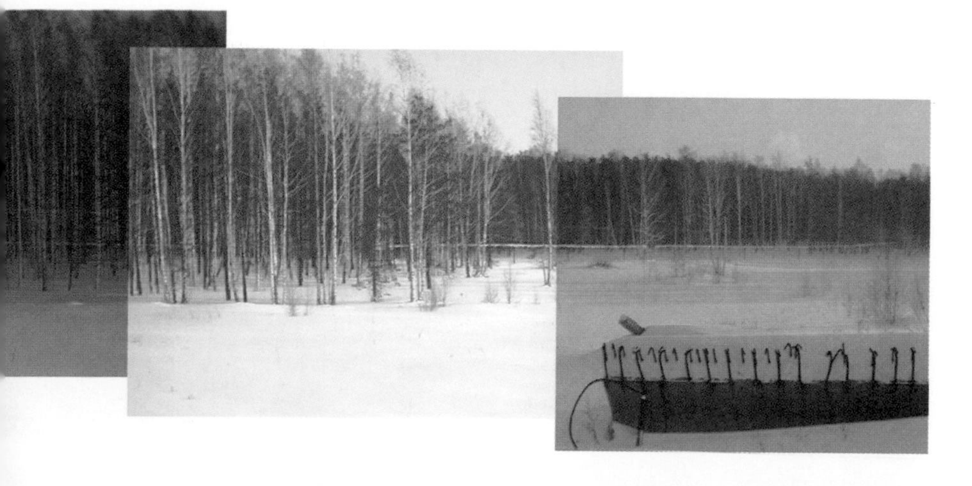

Diane found herself surrounded by nearly 1,000 bones and bone pieces. They were laid out anatomically in nine skeletons. Each numbered skeleton was housed in a spare-no-expense case, open for viewing. She had just met the Romanov royal family—minus two children.

Diane had seen hundreds of skeletons, including too many murder victims to count, but even she was shocked. These bones told a tale of unspeakable brutality. Anyone could see that by the injuries, but Diane's expert eyes saw the horror in all its gory detail.

Bullets had punched perfectly round holes in some of the skulls. Various bones were chipped, splintered, broken, charred, or etched with acid. Some looked as if they had been run over by a truck. Other marks were consistent with blade thrusts. Several faces were smashed to pieces, leaving large gaps where cheeks and eyes and noses had been.

What had these people done to deserve such an ugly fate? Diane wondered. She didn't know much about

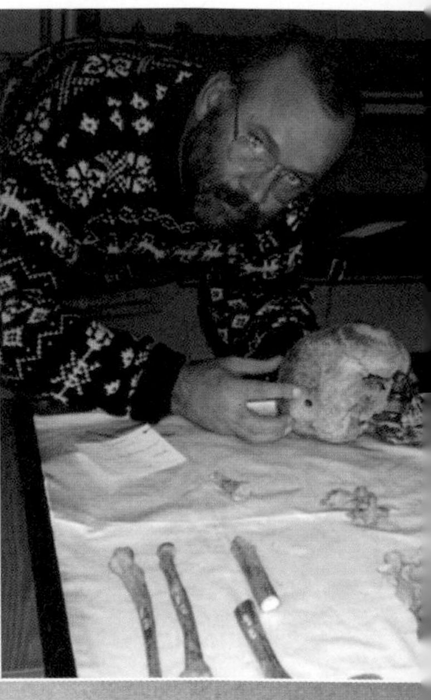

NecroSearch geophysicist Jim Reed points out a bullet hole in the skull of Czar Nicholas II.

The battered skulls of Czarina Alexandra *(left)* and two of her daughters reveal how brutally they were murdered.

Russian history, but one fact was clearly written in these battered bones: The Reds saw the Romanovs as villains, not saints. *Talk about opposing views!*

Diane reached down and picked up a bone from Skeleton No. 2, the family doctor. She held it to her nose and sniffed it. It was an automatic gesture. Diane often used her keen sense of smell to sniff out clues—as she had done at Cher Elder's rocky grave. The skeleton of Dr. Eugene Botkin had an odor the others lacked. Its torso bones were held together by a big lump of adipocere, a waxy substance that forms when body fat meets water in a grave. Its odor is distinct, but the scent changes over time. Surprised the adipocere had lasted 80 years, Diane had sniffed the bone out of curiosity more than anything.

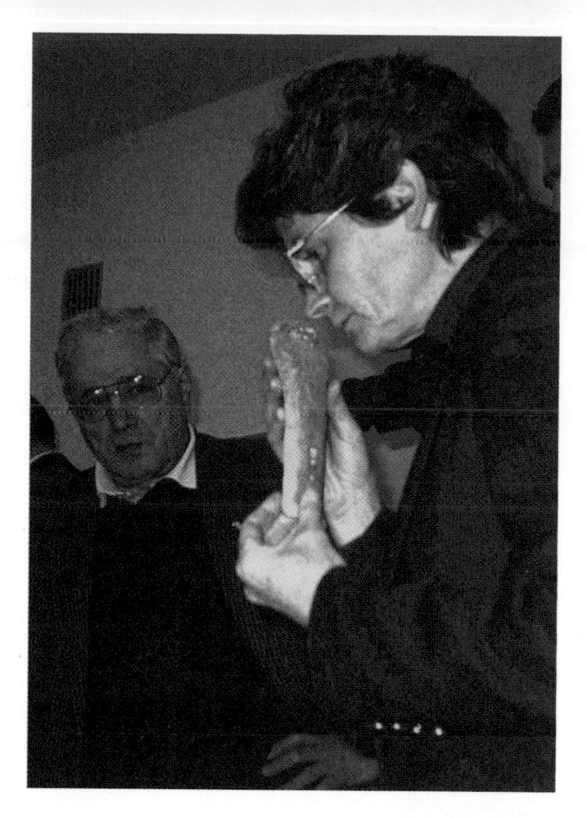

Diane France sniffs a bone from the skeleton of Eugene Botkin, the Romanov family doctor. Although she was only trying to identify certain odors related to decomposition, Diane detected the displeasure of Dr. Avdonin *(left)*.

Suddenly, she felt uneasy. Diane looked up—straight into a pair of hard blue eyes staring at her through large thick glasses. Avdonin wasn't angry; he was *livid.* This time Diane didn't need Galina to translate. She understood Avdonin's body language: *Lady, what do you think you're doing sniffing the Romanovs' doctor?*

Diane hurriedly replaced the bone. *No more being nosy.*

~ Ages Apart

Diane sidled over to the bones of the three grand duchesses, the grown daughters of the czar and czarina. Their beat-up skeletons were numbered 3, 5, and 6. She could tell at a glance that Skeleton No. 3, labeled "Olga," was the oldest. All her limb bones had finished growing.

The growth wasn't so much about getting bigger or longer; it was about fusion. Babies have hundreds of bones, many of which are more like cartilage, which is softer than bone. The bones and cartilage fuse into 206 hard bones by adulthood, a gradual process that

Anastasia strikes a confident pose. She was just 17 years old when soldiers massacred her family.

happens at different times in different parts of the body. Olga's bones had all fused. *She was an adult woman in her 20s,* Diane estimated.

Skeleton No. 5, "Tatiana," was younger than Olga. The limb bones were fused too, but some showed signs of recent growth. There were epiphyseal [i-PI-fi-see-ul] lines visible where two bones had joined. An epiphysis is a small detached bone at the end of a primary bone—a femur or tibia or humerus. Before fusion the epiphysis and main bone are separated by a cushion of cartilage, which crumbles away as the bones grow together. Skeleton No. 5 had just finished that process when she died. *Late teens or maybe 20,* Diane thought.

Age-wise, Skeleton No. 6 fell between No. 3 and No. 5. Diane paused. Something was wrong. The label on No. 6 said "Anastasia," the most famous of the Romanov daughters and the only one Diane could have named before this adventure started. *But wasn't Anastasia the youngest of the four daughters?* Anastasia had been a teenager, which meant these bones didn't add up. Skeleton No. 6 was clearly older than No. 5. So, if any skeleton was Anastasia, it was No. 5, the youngest skeleton.

Diane pulled Jim and Tony aside. She pointed to No. 6 and asked Tony, the other anthropologist, "Does this skeleton look like a teenager to you?"

"No way," he agreed. "Bill Maples thought it was Tatiana, the 21-year-old. Anastasia was barely 17."

"What about No. 5, the youngest skeleton?" Diane asked.

"Bill said it was most likely Marie, the 19-year-old."

How Grown Are the Bones?

Pinpointing the precise age of a skeleton is difficult but not impossible. From studying thousands of skeletons of known age, anthropologists know that the femur of a girl grows from birth to adulthood as shown below.

Because the femurs of Romanov Skeleton No. 3 lacked fusion lines, Russian and American scientists alike identified it as belonging to 23-year-old Olga *(at right in photo)*. Skeleton No. 6, though labeled "Anastasia," had fully fused limb bones. To Diane and her colleague Bill Maples, this made it likely that No. 6 was 21-year-old Tatiana *(left in photo)*.

The Russians disagreed. They had identified Skeleton No. 5, not 6, as Tatiana. But to Diane's practiced eye, the bone growth of No. 5 signaled a younger age—someone in her late teens, such as 19-year-old Marie. Following this reasoning, Diane concluded that none of the three skeletons were young enough to be Anastasia *(center in photo)*. The whereabouts of the youngest Romanov daughter, Diane believes, remain a mystery.

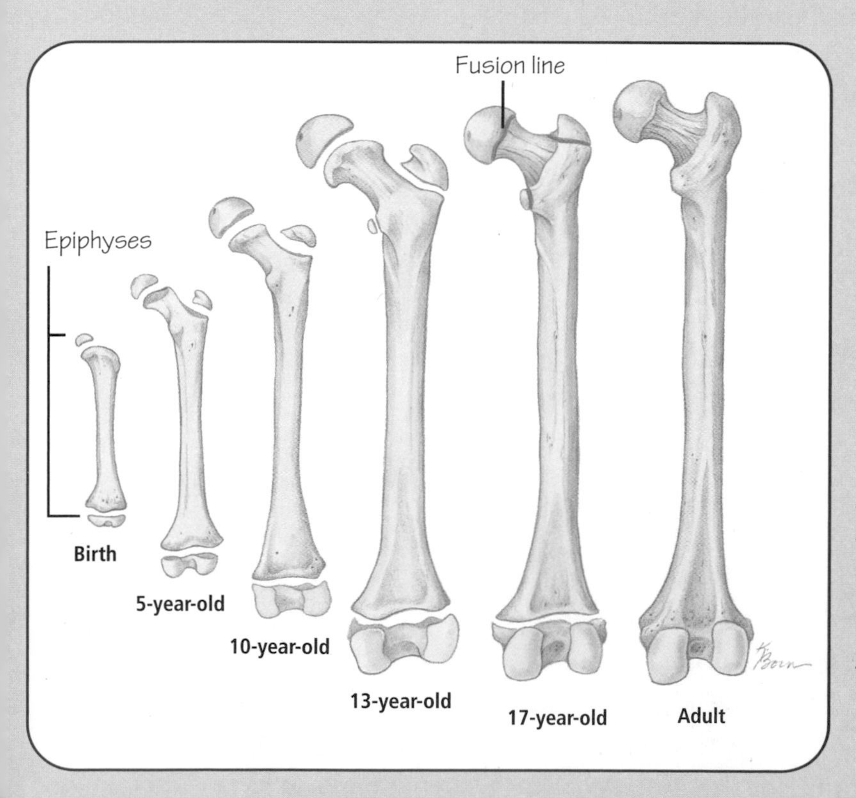

Bones start out in pieces, then fuse together as they grow. The smaller end pieces, called epiphyses (e-PI-fi-seez), are connected to the primary bone by cartilage. As the body ages, osteoblasts gradually ossify that cartilage—that is, they turn it into bone. Some bones, including the head of the femur, fuse as early as the midteens. Others—the vertebrae and collarbone, for example—may not fuse until the early 20s.

Bill Maples had also disagreed with the height estimates. His figures were based on measurements of the long bones of the three skeletons. These figures excluded Anastasia, who was much shorter than her sisters. The daughters' exact heights are unknown, but their relative heights are visible in photographs.

Russian scientists were convinced that Marie was missing, but American scientists were sure that the missing skeleton belonged to Anastasia. Diane found scientific errors like these offensive—wrongs that absolutely had to be righted—but she reminded herself why she was there. It wasn't to enter a Cold War between Russian and American scientists. It was to find the bodies of two missing children, whatever their names. She, Tony, and Jim agreed to focus on their NecroSearch mission.

~ *A Hot War*

The next day the group met to sign an agreement. Avdonin's office was in a museum full of posters, uniforms, and other communist artifacts. The room was chilly—no surprise there—but clean and bright. To cut costs, little-used halls and rooms remained unheated, unlit, and unclean. To Diane's dismay, that included the ladies' room.

Avdonin sat behind a large desk with his wife Galina at his side. He presented the Americans with a paper in Russian, and Peter began to translate it. He didn't get past the first sentence. The document stated that NecroSearch was to look for Alexis and Marie.

> *"Nyet!"* cried Avdonin. "It is Marie! Fifty scientists are not wrong about this. It is *you* who are wrong!"

Diane felt hot. *Oh no,* she thought. *Here we go.* She calmly explained why she couldn't sign the paper: She believed the missing daughter was almost certainly Anastasia, not Marie. As she began to talk about growth plates, Avdonin cocked his head and his face flushed red. He began yelling in Russian.

Galina tried to keep up with the translation. "Who do you think you are? You Americans, you come over here. You take one look at the bones, and you contradict us. Three of the top Russian scientists say it is Marie who is missing. They are right!"

Diane said, "I still can't sign it." She had built a solid reputation for being accurate. *No way am I going to risk losing that.*

~ Yell All You Want

Avdonin exploded. "We have 10 scientists who say you are wrong. They have studied these bones for months! How can you make such a decision in two minutes?"

Galina looked worried. She said to Diane, "Please! Can't you see what you're doing to him? Please, just sign the paper."

Diane shook her head, but she had an idea. "What if we just change the wording?" she proposed. "We could say that we're looking for Alexis and one of the Romanov daughters without naming her."

"Nyet!" cried Avdonin. "It is Marie! Fifty scientists are not wrong about this. It is *you* who are wrong!"

Diane paused. Avdonin had a point. She hadn't seen all of the evidence. Anastasia's gender, age, and height weren't the only clues. Russian scientists relied heavily on photo superimposition. Basically, a computer combined a photo of a face with a photo of a skull. Then, the scientists compared detailed features and measurements, point by point, to see if they matched. For example, the skull of Skeleton No. 3 had a bulging forehead; so did Olga, according to photos. The Russian team concluded that the skull of Skeleton No. 6 did not match photos of Tatiana or Marie. That left Anastasia.

Diane thought the technique was shaky, especially as she pictured the smashed faces of the grand duchesses. Tiny pieces of skull had to be glued together, like shattered pieces of pottery or more like Humpty Dumpty. So many pieces were missing or damaged. The reconstruction would have holes and gaps and jagged edges. The glue itself could skew the positioning. Even with a perfect skull and an excellent photograph, Diane didn't think photo superimposition alone was reliable. In her view it

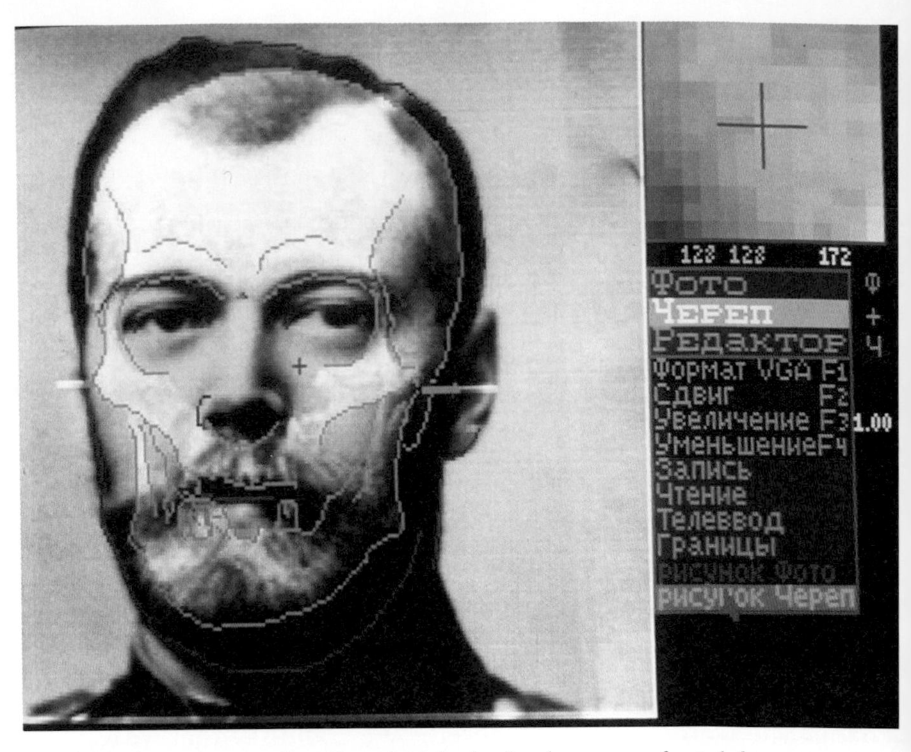

Using a computer, Russian scientists combined an image of the czar's skull with a photo of his face. Then, they looked for matching features—the shape of the forehead, the width of the cheeks, the length of the face, and so on. Most of the features matched.

could rule out people who clearly lacked certain facial features. But without other clues it couldn't positively match a skull with an individual. For example, only someone with a bulging forehead could have belonged to skull No. 3, but a lot of people have bulging foreheads. Olga's identification was solid because of the other data—age, height, gender, tooth fillings, and so on.

Even so, as a scientist, Diane felt she had to consider all the evidence. She asked Avdonin, "Could we see the data?"

"Da," he agreed, to Diane's surprise. "But everything is in Moscow. Now, you must sign the paper. Later, you can look at the data."

Diane shook her head. She couldn't do it. *It would be like lying,* she thought.

Avdonin lost control, ranting and raving. Peter, like Galina, worried that the geologist would have a heart attack or a stroke. He understood that Avdonin faced pressures on every front.

The geologist had to deal with a communist government in Moscow that wanted to bury and forget the Romanovs. Members of the Russian Orthodox Church wanted to canonize the royals as saints, which would mean the skeletons couldn't be buried. Also, there were the feelings of relatives to consider—Romanov cousins, nieces, and nephews. On top of that, dozens of elderly impostors claimed to be Alexis or Anastasia, creating more problems. Now, here were more arrogant Americans saying that Russian scientists didn't know what they were doing.

As Avdonin screamed at Diane, Jim tilted his computer screen toward her. *"Pravda,"* he had typed—Russian for "truth." Diane smiled. Jim was right. To her this wasn't about politics or religion or emotions. This was about the scientific search for truth. She had dedicated her life to that search, and she wasn't about to stop now.

The "hot war" continued. By late afternoon, Diane noted, Avdonin was up to 200 Russian scientists who knew more than she did. She had had enough. She was tired of being yelled at. After talking to the group, she told Avdonin, "We're going home."

As the Americans began to leave, Avdonin caved. He invited them to stay, promising to change the document just as Diane had suggested.

Yeah, four hours ago, Diane thought.

~ *Where Is Anastasia?*

That night the Russians and Americans celebrated together. Jim played boogie-woogie on the piano, and they all laughed, sang, and told stories. Diane was baffled. This guy Avdonin had been unbelievably rude and insulting to all of them. She even told him so to his face. Now, he was acting like their new best friend.

The next day at the airport Avdonin gave her a signed copy of a book he had written. He told her, "You're beautiful when you're angry."

In the interest of Russian-American cooperation, she forced a smile. During the long flight back to the States, Diane and Jim called the NecroSearch team.

"Avdonin is rude and a real pain to work with," Diane reported. "He might not ask us back to do our job. But if he does, it'll be worth it."

As Diane suspected, NecroSearch wasn't invited back that spring, probably because Avdonin was still angry. Even though two bodies were missing, the Romanov funeral took place that

Eighty years after the Romanovs met their fate in a dingy St. Petersburg basement, Russian president Boris Yeltsin held a state funeral for them in this grand hall. Russian citizens turned out *(inset)* to honor their memory.

summer on July 17, 1998, in St. Petersburg, Russia. The date was exactly 80 years after the mass execution. Skeleton No. 6—still labeled "Anastasia"—was put to rest. So, too, declared Russian President Boris Yeltsin, was "one of the most shameful pages in our history."

In February 2000, Diane and Russian scientist Sergei Nikitin presented the case at the American Academy of Forensic Sciences conference in Reno, Nevada. Diane reviewed the evidence that convinced American scientists that Skeleton No. 6 was not

Anastasia. Sergei explained why photo superimposition convinced Russian scientists that it was. He then presented a facial reconstruction of "Anastasia"—a head sculpted on a cast of the pieced-together skull. The room was packed. Though long dead—and now formally buried—the Russian royal family still drew quite a crowd.

Later that year Avdonin finally invited Jim Reed to return with his computer program and geophysical instruments. Bags packed, instruments tuned, Jim couldn't wait. Then, sadly, Avdonin suffered a stroke, and the trip was canceled. Peter has continued to search on his own time and money, but so far this royal fairy tale has no ending. As of this writing, the bones of the two Romanov children have never been found.

One summer day in 1995,

Diane stared into the empty eye sockets of

Jesse James,

the notorious American outlaw.

Or did she?

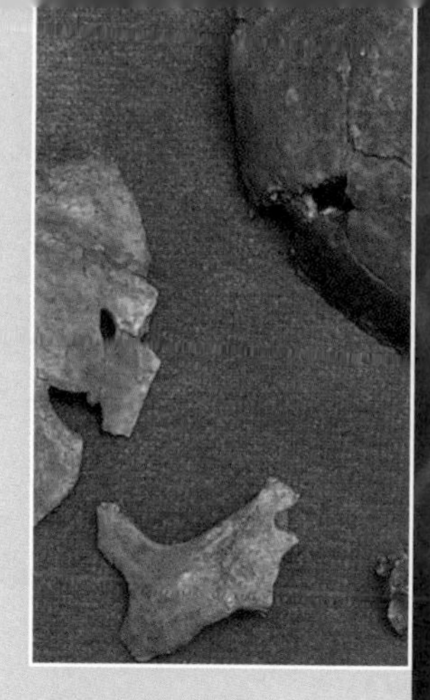

8

CASTING CHARACTERS FROM HISTORY

The bones of the rich and famous weren't limited to Russian royalty. Starting in the early 1990s, Diane opened her casting lab door to a parade of dead but colorful characters from history: early settlers, soldiers, sailors, heroes, and villains. People hired her to make plastic casts of historic bones for a range of reasons. Archaeologists and other anthropologists could continue to study the casts after reburying the remains. Artists used Diane's skull casts to sculpt clay reconstructions of heads. Museums displayed the casts in exhibits.

Casting was Diane's official job. Unofficially, no physical anthropologist could resist examining bones for clues to a person's life. Every skeleton—historic or otherwise—had a fascinating tale to tell.

One summer day in 1995, Diane stared into the empty eye sockets of Jesse James, the notorious American outlaw. Or did she?

In 1864, Jesse James *(opposite)* belonged to a pro-Confederate band of guerrillas. After the Civil War, he headed west and formed a gang of Wild West outlaws. Diane created a cast from Jesse's broken skull *(above)*.

~ *Under the Name of Jesse James*

The fragile skull and other remains came from a double grave in the James family plot at Mt. Olivet Cemetery in Kearney, Missouri. The tombstone read "Jesse Woodson James, September 5, 1847, assassinated April 3, 1882."

But was the corpse under the marker really him? Jim Starrs, a professor of law and forensic science, paid to have the body exhumed in order to solve that history mystery.

According to old rumors, another man had been buried in Jesse's place to allow the outlaw to escape. Jesse James supposedly moved —to Texas, some said—and fathered children under a different name. And now a number of people were claiming to be descendants of those alleged kids. If true, these people would be the great-grandchildren of a robber and murderer. At face value, it wasn't much to brag about, but some Americans treat outlaws as heroes.

Jesse James looks the dapper gent in this photo from 1882, the year he was shot. Rumors arose quickly that he had survived and that someone else was buried in his grave. Forensic anthropologists examined that skeleton and concluded it was likely Jesse.

After hiring a science team to study the bones, Jim would eventually rebury them. He asked Diane to make a cast of the skull in order to preserve the information it contained for future research. Diane agreed, but she ignored the "Is He or Isn't He?" debate. Jesse's many victims had been dead and buried for more than a century. She felt it was far more important to help the families of recent murder victims. Still, as she turned the skull around in her hands, she couldn't help looking for the famous bullet hole.

The story of the outlaw's death was legend. On April 3, 1882, Jesse James, under the alias Thomas Howard, was hiding from Pinkerton Agency detectives on a farm near St. Joseph, Missouri. Bob Ford, a former member of the bank- and train-robbing James gang, wanted the $10,000 reward for Jesse's capture. And he knew just where to find the outlaw. Catching a gunslinger dead was far safer than catching him alive, so as Jesse James dusted a picture

An 1882 drawing demonizes Bob Ford for shooting Jesse James in the back of the head. In this homey scene, James seems the hero and Ford the coward.

on the wall, Bob shot him in the back of the head. But instead of a reward, Bob Ford earned a murder conviction.

Newspapers reported that Jesse James, a.k.a. Thomas Howard, was dead at the age of 34. Someone even wrote a song about it: "That dirty little coward that shot Mr. Howard / Has laid poor Jesse in his grave."

~ *Drawing Conclusions—Sort of*

The bullet hole was hard to find because, after more than a century underground, the skull was a mess. It was cracked in places and had pieces missing. Diane noticed a possible gunshot wound in the right base of the skull. The hole was incomplete—part of the edge was missing—but the rim was beveled. Beveling happens when a bullet hits bone, makes a neat, round entry hole, but then breaks off a cone-shaped piece on the exit side of the bone. Of course, lots of objects—an umbrella tip even—can bevel bones. Diane didn't really think someone poked Jesse's head with an umbrella, but she couldn't say for sure that a bullet made the hole either.

The report of the science team, which included forensic anthropologist Mike Finnegan, said: "The likely entrance wound . . . appears to be in the base of the man's skull." "Likely" and "appears"—Diane approved of those weasel words, as she liked to call them. Inexact adjectives like "probably," "possibly," and "consistent with" are vital to science reports. They help scientists avoid saying anything definite that the evidence doesn't support. In Jesse's case, saying "That's a bullet hole" would be more speculation than fact.

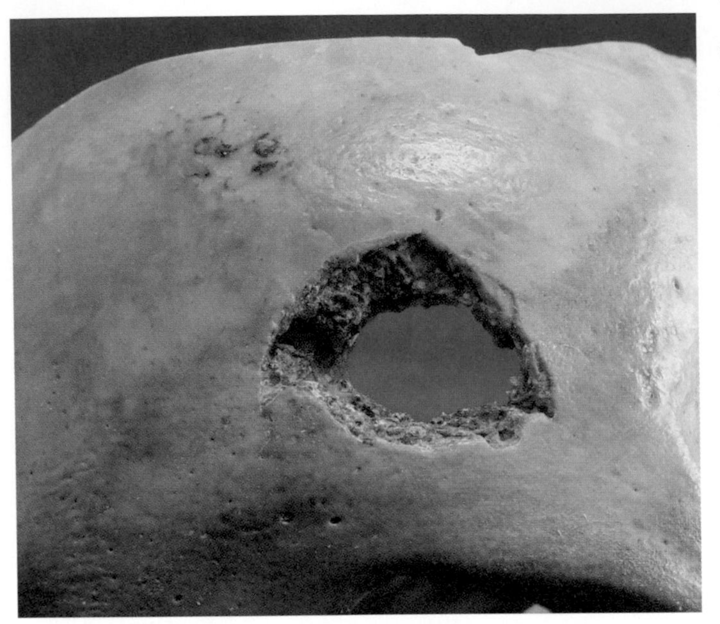

A bullet made the hole in this bone. But did it go in or come out this side? The answer lies in the beveled, or angled, edges surrounding the hole: The bullet entered from the other side, then chipped off a cone-shaped piece of bone as it exited the side shown above.

Other skeletal evidence in the Starrs report was consistent with—but also didn't prove—an identification of Jesse James. Bone measurements and observations confirmed the victim's age (30s), height (5′ 8″ to 5′ 10″), and sex (male). There was a bullet in the right rib cage—consistent with a nonfatal wound that Jesse James had suffered during the Civil War.

Scientists tested the teeth, hair, and long bones for the presence of mitochondrial DNA (mtDNA), which is genetic material passed down only through the maternal (mother's) line. The mtDNA in some of the bone tissue matched the mtDNA from two living descendants of Jesse's sister, Susan. But even that evidence wasn't foolproof. As the report noted, "There is always the possibility (however remote) that the remains are from a different maternal relative . . . or from an unrelated person with the same mtDNA sequence."

The report concluded with more weasel words: "There is no scientific basis whatsoever for doubting that the exhumed remains are those of Jesse James." The evidence didn't prove that the skeleton belonged to Jesse James, but it couldn't prove that it belonged to someone else either.

~ The Hunley Boys

"All of these are *Hunley* boys," Diane told two visitors to her casting lab one summer day in 2003. She pointed to the casts of eight well-preserved skulls lining a wooden shelf.

The *H. L. Hunley* was the first submarine to sink an enemy ship during a war—the American Civil War. The Confederate vessel predated electric power, motorized engines, and long-range torpedoes. It was powered by human muscle. The airtight iron tube was just long enough to fit a commander and seven sailors seated shoulder to shoulder the length of a wooden bench. The sailors worked by candlelight to rotate a long crank that turned the submarine's propeller.

On February 17, 1864, the commander, Lieutenant George Dixon, age 24, steered his submarine toward the *Housatonic*. The Union warship was blockading Charleston Harbor in South Carolina, cutting off the Confederate line of supplies.

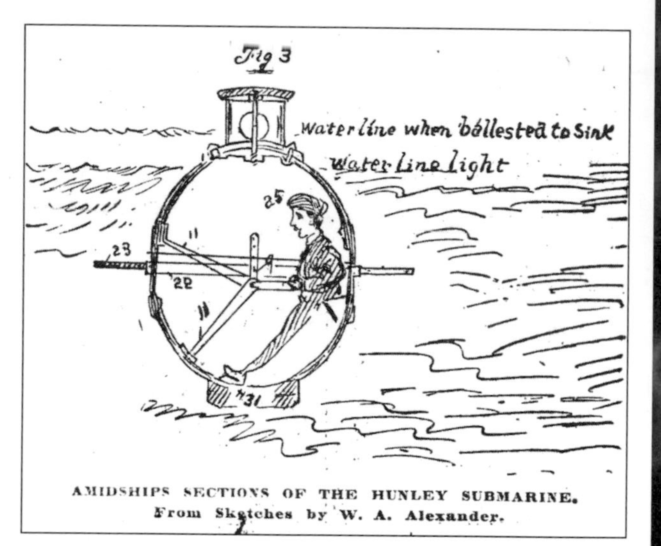

To enter the submarine *Hunley (above)*, each sailor squeezed through a hatch about the size of a modern car tire. With no room to stand or move, the crewmen sat for hours, turning a crank that propelled the warship underwater.

An 1863 painting shows the *Hunley* at its dry dock. During test runs, the submarine sank twice, drowning 13 men—including its namesake, inventor Horace Hunley. Despite the danger, Lt. George Dixon quickly recruited a third and final crew.

Upon reaching the target, the submarine rammed the Union ship, inserting a torpedo in the hull, and then backed away. The torpedo exploded, sinking the *Housatonic*. The *Hunley* escaped unharmed. But then, minutes later, the submarine mysteriously disappeared before it could reach shore. All hands were lost.

In 1995, shipwreck searchers spotted the *Hunley* on the bottom of the harbor. In 2000, a team of engineers carefully lifted the submarine out of the water in a giant truss. The ship was taken to the Warren Lasch Conservation Center, a laboratory at the old Charleston Navy Base. There an archaeology team led by Maria Jacobsen began excavation and found eight surprisingly well-preserved skeletons inside. Maria hired Diane to make casts of the sailors' bones before a burial in April 2004.

Diane picked up one of the skull casts in her lab and held it up for her visitors to see. She pointed out two rounded gaps in the teeth on either side of the jaw.

"This guy smoked a pipe," she explained. "He started on one side and then, when the pipe ground down the teeth, he had to switch to the other."

Next Diane picked up a shiny replica of a $20 gold coin found by Maria Jacobsen. It was warped and twisted. The coin had led Maria to identify George Dixon as the submarine commander.

One hundred and thirty-six years after sinking to the bottom of Charleston Bay, the *Hunley* emerges from its watery grave in a truss—a metal cage that acts like a sling.

Crouched in the *Hunley*, archaeologist Maria Jacobsen marvels at a prize find—a gold coin given to Lt. Dixon by his girlfriend, Queenie Bennett. Descendants of Queenie and of three crewmen attended the funeral of the *Hunley* boys in April 2004.

A Dixon family tale related how a coin—a gift from the lieutenant's girlfriend, Queenie—had stopped a Yankee bullet from killing him. From that day on, Lt. Dixon had carried the mangled coin in his pocket for good luck. To Maria's surprise, the story turned out to be true. The coin, issued in 1860, was inscribed "Shiloh. April 6th, 1862. My life Preserver. G.E.D."

Diane pointed to an area of reactive bone on the femur. Reactive bone is a rough patch where there was an injury—heavy bruising from a bullet, for example. The injury stimulates osteoblasts to grow new bone. On Dixon's skeleton the reactive bone was near the hip joint, right at pocket height.

The *Hunley* recovery team planned to display Diane's casts in a museum that was about to be built. But did the casts belong in a public exhibit? Some people objected to displaying soldiers' remains—even though they were plastic replicas. Diane saw nothing wrong with it. In fact, she had no qualms about showing the real bones.

"If people want to really learn about the soldiers," she said, "you have to show the bones. The bone is a record of a person's life, especially the last part—the circumstances of death."

How and why did the *Hunley* sink? What can we learn about the lives—and deaths—of its ill-fated crew? Scientists will likely be investigating the answers for years to come, thanks in part to Diane's casts.

~ A Living Cast

Diane's favorite "casting character" of all time wasn't dead, famous, or even human. In March 1997, the National Zoo in Washington, D.C., asked her to make casts of the paw and tongue of a living tiger for an exhibit. Tiger paws have sharp claws, and the tongue is sandwiched between four long, meat-ripping canine teeth. But Diane wasn't worried about her safety, since the big cat would be tranquilized. She was more worried about the tiger. She didn't want to harm the animal in any way. She didn't know—yet—how to cast the body part of a living animal, especially one so large and . . . *predatory.*

Diane stood in front of a mirror and stuck out her tongue. It looked like a rosy pink welcome mat to her delicate insides. Her usual rubber mold formula was out of the question—too toxic. She wondered, *What is safe to swallow, fast and easy to apply, and hardens quickly into a firm mold?*

One of her horse-riding pals, a veterinarian named Mary Wright, had the answer: alginate. Mary told Diane that alginate was a pink gummy material used to make molds of teeth. If it was safe for pets, it should be safe for a tiger.

And people, too, Diane thought.

At her lab she mixed the alginate powder with water to form a paste. Standing in front of a mirror, she scooped up a blob and spread it on her tongue. She pressed it down with her fingers to get out any air bubbles. *Not bad,* she thought. *Tastes just like spearmint.* She decided to make an extra-thick mold to make it easier to peel off. So, she added another gloopy blob—and another.

With her thickened tongue hanging out, Diane padded around the lab doing odds and ends while the material set. After about 10 minutes, she peeled off the alginate slowly and carefully. It was fragile—bendable and easy to tear—but it made an excellent impression of her tongue.

Next Diane borrowed Mary's cat, Whites, for a dress rehearsal of the big event. Mary told Diane that cats vent body heat through the tongue. To avoid overheating the animal, she recommended using cold water—and working quickly.

When Mary tranquilized her cat for its regular teeth cleaning, Diane made an impression of the little pink tongue. She had to work around an oxygen tube in the cat's mouth. But, once again, she was able to produce an amazingly detailed impression. It included all the spines that made the cat's tongue so rough. Cats need the stiff bristles—called filiform papillae—to clean their fur. Diane wondered what the tiger tongue would feel like—a scrub brush?

Diane fine-tuned her flexible casting skills with the help of Whites the cat.

~ A Tiger by the Tongue

At the zoo an animal dentist wheeled in a tranquilized Bengal tiger, a creature twice as massive as a man. It probably weighed more than 300 pounds. Its muscular, orange-and-black-striped body spilled over the sides of the operating cart. *What an impossibly beautiful animal,* Diane thought.

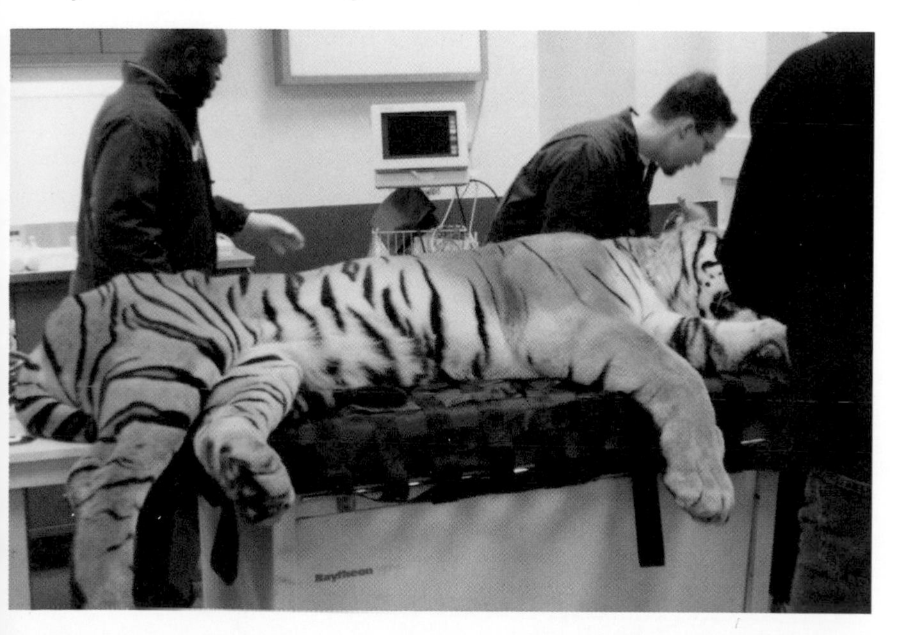

National Zoo technicians work quickly to help Diane make alginate (gummy) casts of the paw and tongue of a sedated Bengal tiger.

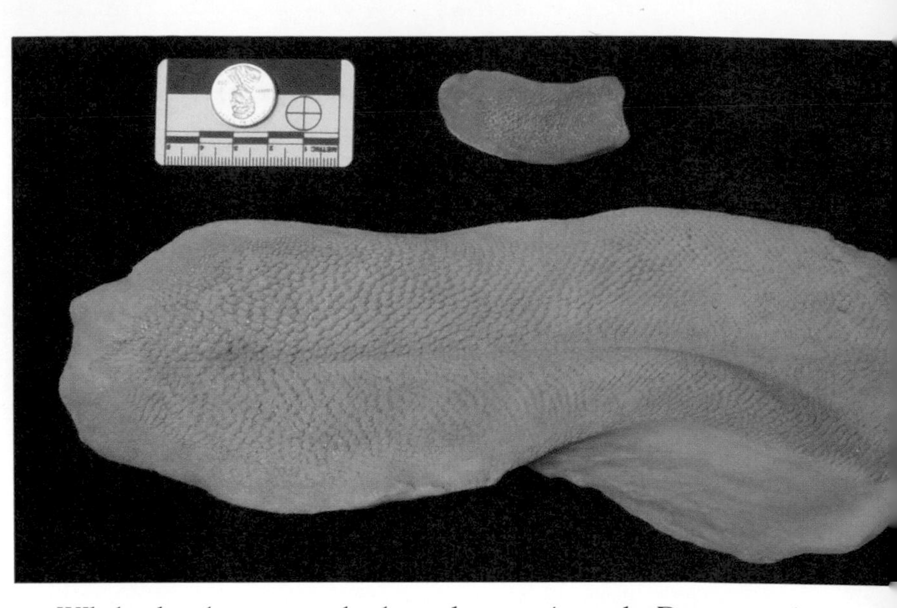

Diane's cast of the nine-inch-long tiger tongue dwarfs that of Whites the cat. (A penny shows the scale.) The papillae (bristles) on the tiger tongue are rough enough to lick raw meat right off the bone.

While the dentist worked on the tiger's teeth, Diane made an alginate impression of one of its huge paws. Then, the dentist used two cupped hands to pull the tongue out of the big cat's mouth. The heavily muscled organ was almost as long as her foot. With the dentist still holding the tiger tongue in both hands, Diane quickly went to work. The pink surface was rough, like sandpaper, but floppy and wet. She coated the top with alginate and waited for it to set. Her 15 minutes were almost up. The tiger, so calm and soft, would soon wake up and be dangerous again. Slowly and carefully, Diane peeled off the soft impression to avoid tearing it.

Please be rude: Invited to do so, a National Zoo visitor compares Diane's tiger tongue cast to her own tongue in a mirror.

Diane made a rubber silicone mold of the impression. Then, she made a sturdy wax cast of the mold and sent it to a metalworks company. The foundry workers coated the wax mold with plaster and put it in an industrial oven. The wax melted off, leaving a plaster shell. The workers poured melted bronze into the shell, and the metal cooled and hardened into a bronze tiger tongue.

Diane's casts of the tongue and paw allowed zoo visitors to share her amazing experience. People felt the roughness of the big cat's tongue and spread a hand inside the palm of its giant paw. *This,* she decided, *was one of the coolest things I have ever done.* She savored the moment, knowing that far graver matters were always just a phone call away.

The entire nation was stunned—
and frightened.
Would there be
more ***attacks?***

Who killed all these people—
and ***why?***

9

Disaster Is One Call Away

One summer afternoon in 1997, Diane was sitting in her car when her cell phone rang. There was a time in her life when answering the phone was no big deal, like turning on the television. But now, as a forensic anthropologist, she was always one call away from someone's tragedy.

"A hiker found an unknown human skeleton. Can you identify it?"

"The CSU lab has just received a murder victim. Please examine it for evidence."

"The sheriff has a hot tip on the whereabouts of a missing body. Is a NecroSearch team available?"

Of course, the call could just be Joy or LuAnn inviting her on a shopping spree. Or, with any luck, the call was another casting job. This time, though, it turned out to be none of the above. It was the worst news possible: a mass fatality.

Diane was a member of DMORT—Disaster Mortuary Operational Response Teams. This U.S. government organization sends armies of skilled workers to disasters that are too big for local personnel to handle. The teams include medical examiners, forensic dentists, DNA experts, morticians, psychologists, detectives, criminalists, excavating machine operators, and medical record keepers. Computer experts organize everyone and everything into databases.

A Coast Guard crew patrols New York Harbor *(opposite)* to keep boats away from the site of a mass fatality—the toppled World Trade Center towers—on September 11, 2001. Diane *(above)* investigates a crime scene.

Diane was one of the first anthropologists to join the group when it was formed in the early 1990s.

The DMORT team leader filled her in on the details: About two hours after midnight, Korean Air flight 801 had crashed into a hillside jungle on Guam, a tropical island in the South Pacific Ocean. There were roughly 235 dead. Diane was booked on a plane out of Denver to leave in about four hours.

~ *Guam*

Diane drove home, called her husband, Cal, and packed a bag, making sure to toss in her DMORT identification card. She knew that a tough job awaited her on the other side of the world. A DMORT shift meant two straight weeks of 13-hour days, after which workers were advised to go home so they wouldn't burn out.

In one sense, Diane was relieved to leave Colorado for a while. She and Cal hadn't been getting along well. Her career was really taking off, yet he still treated her like a junior partner. She hated it when he burst out in anger, for no apparent reason, and yelled at her. Her marriage was in trouble, and she wasn't sure what to do about it. The 31-hour plane ride from Detroit to Japan to Guam would give her time to rest and think.

The plane landed in Guam at night during a rainstorm. The next day, after just a few hours of sleep, the DMORT team went

More than 200 people perished in the 1997 crash of this Korean Air jumbo jet on the Pacific island of Guam. Many of the survivors were seated in the tail section, which held together on impact.

to the crash site. The Boeing 747 had gone down a few miles short of the runway, breaking into four large pieces that were still smoking. After seeing the debris, Diane was stunned to learn that 32 of the 254 people on board had survived the impact. Three survivors died at the hospital. But a flight attendant walked away with only cuts and bruises after being flung clear of the aircraft. Others, including a young girl, were pulled out alive from the larger sections—the tail and first-class cabin.

The victims' relatives, several hundred in number, kept pouring into the airport. Though news of survivors gave them a glimmer of hope, it also left people wondering: Were their loved ones dead or, miraculously, alive? A survivor list was posted, but many relatives clung to the faint possibility that more names would be added.

Diane knew that the other list, the names of the dead, would take much longer to compile.

~ A Life-Changing Decision

U.S. servicemen and other rescuers brought the remains in refrigerated trucks to a giant airline hangar, the temporary DMORT morgue. As usual, Diane prepared a mental box to store her emotions, tied it shut with ribbon, and added it to the sagging shelf in her mind. She sometimes had trouble sleeping and eating after a tough case, but who didn't? The bottom line was the box trick seemed to work; she always got the job done.

Each day of her DMORT shift, Diane woke up at dawn and worked until dusk. At one point she heard a faint crackling sound, like puffed cereal in milk, coming from a body bag. She could see the plastic moving ever so slightly.

Diane paused. The tropical air felt like a hot, wet towel plastered to her face, but she was used to tough field conditions. The bigger challenge of this climate was decomposition. Decay happens much faster in a warm and moist environment. With decomposition, Diane knew, came decomposers. Inside the crackling bag on her table, experience told her, was a mass of maggots doing what came naturally to them: consuming tissue.

Bugs were an unpleasant part of the job, but Diane had steeled herself against them long ago. This time she hesitated for a different reason. That very instant, her hand poised on the bag's zipper, Diane made a life-changing decision: She decided to divorce Cal. *I can't do both,* she thought. *I can't work these mass fatality cases and then go home to unpleasantness of a different kind.* Grisly as her job was, she couldn't imagine doing anything else. But to keep doing it, she knew she needed peace at home.

Back in Colorado, Diane told Cal, "We need to talk."

Before she could say another word, he asked, "Are you moving out, or am I?"

To Diane's relief, Cal had been thinking the same thing. She and her dog, Moki, found another place to live.

~ *Taking a Break*

On October 31, 1999, Halloween morning, another jumbo jet crashed. For no apparent reason, Egypt Air 990 made a sudden, steep dive into the Atlantic Ocean shortly after takeoff from JFK airport in New York City. This time the relatives were told right away there was no possibility of survivors.

For the first time in her life, Diane said "No" to a call for help.

When the DMORT call came, Diane dutifully packed her bag, which now included an official uniform—khaki military pants and a dark green shirt. She flew east to work the first two-week shift after the accident.

Diane put on her Dr. France mask, prepared a ribboned box, and went to work. While pushing herself to the limit, she always felt an adrenaline rush—a burst of energy that helped her get through those 13-hour days. But then, all of a sudden, there was nothing to do. Bad weather scrubbed the ocean recovery effort, so Diane headed back to Colorado.

Days later, after a routine doctor's visit, she was informed she had cancer in her thyroid, a gland in her throat. Diane had had no symptoms of the disease, which is treatable with surgery followed by doses of a chemical that kills off the cancer cells.

After a successful operation in January 2000, she went home to recover, throat bandaged.

Within the hour, the phone rang. It was Frank Saul, a DMORT team leader. The Egypt Air operation was back on, and they needed her right away. Diane couldn't believe it. She hadn't even been home for a full hour. Frank insisted it was urgent: Could she come the following day, then, or the day after that?

Riding Handy the horse fast across the Colorado wilderness was a great way for Diane to unwind.

In the next few weeks, Diane would have to undergo radiation treatment. She was told it would wipe her out physically. She would be too weak to walk across the room. There was no way she could fly to New York and work 13-hour days.

For the first time in her life, Diane said "No" to a call for help. She said it because she had no other choice, but those two little letters were like a knock on the head. *Why didn't I think of that before?* She suddenly realized she didn't have to work the mass fatalities. Other anthropologists could shoulder that burden for a while.

The realization freed Diane to take a much-needed, yearlong break from mass fatality cases. Finally, cancer free, she set off on a restful, 100-mile solo trek in England. When she returned, she took over as president of NecroSearch and continued working murder cases. Then, in 2001, she joined the biggest forensic investigation in history.

~ 9/11

Diane woke around dawn on September 11 and made a pot of coffee. She flipped on the TV to see how a Denver Broncos football player was doing after breaking his leg, badly, in two places. She wasn't an avid fan—like her dad—but this was a big story in Colorado. As Diane was about to hop in the shower, she heard news that kicked this "big story"—and every other story—

off the air for days. She watched, wide-eyed, as a news videotape showed a jumbo jet crashing into the north tower of the World Trade Center in New York City. The tower was still on fire when another jet flew into the south tower.

Diane didn't even wait for the phone call. There had to be thousands of people in those office buildings, and she knew that DMORT would need her. Diane started packing her bag in case she pulled the first shift. Less than an hour later, a third hijacked jet crashed into the Pentagon building in Arlington, Virginia. Twenty minutes after that, incredibly, the south tower of the World Trade Center collapsed in a black cloud of dust and debris. That tragedy was followed by a fourth hijacked jet crashing in rural Pennsylvania. Then, around 8:30 A.M. on Diane's clock, the World Trade Center's north tower collapsed.

The entire nation was stunned—and frightened. *Would there be more attacks? Who killed all these people—and why?*

Two weeks later, in late September, the DMORT call finally came. Diane flew to New York City to work the second shift after the terrorist attacks. She took up a post on Staten Island, where tons of recovered material from the World Trade Center site were being sifted and sorted. Recovery teams raked the rubble, sometimes on their hands and knees, removing anything that looked like evidence. Diane's job was to identify and sort human tissue, such as teeth or pieces of bone, in the rubble.

All the remains were taken to 18 refrigerated trucks outside the medical examiner's office in Manhattan, the heart of the city. In the first year after the disaster, DNA tests and other forensic techniques would positively identify about one-third of the nearly 3,000 victims.

FORT COLLINS
COLORADOAN
www.coloradoan.com

Today's weather
Mostly sunny and warmer, clear skies tonight.
High: 76
Low: 50

error!
D.C., New York targeted

Smoke rises from the Pentagon Tuesday after the building took a direct hit from an aircraft in a terrorist attack.

World Trade towers, Pentagon hit in apparent terrorist attack

By JERRY SCHWARTZ
The Associated Press

NEW YORK — In one of the most audacious attacks ever against the United States, terrorists crashed two airliners into the World Trade Center in a deadly series of blows Tuesday that brought down the twin 110-story towers. A plane also slammed into the Pentagon as the government itself came under attack.

Thousands could be dead or injured, a high-ranking New York City police official said, speaking on condition of anonymity.

Authorities had been trying to evacuate those who work in the twin towers when the glass-and- total of 156 people. But the airline later said that was unconfirmed. Two United airliners with a total of 110 aboard also crashed — one outside Pittsburgh, the other in a location not immediately identified.

"This is perhaps the most audacious terrorist attack that's ever taken place in the world," said Chris Yates, an aviation expert at Jane's Transport in London. "It takes a logistics operation from the terror group involved that is second to none. Only a very small handful of terror groups is on that list. ... I would name at the top of the list Osama bin Laden."

President Bush ordered a full-scale investigation to "hunt down I don't think that I overstate it," said Sen. Chuck Hagel, R-Neb.

In June, a U.S. judge had set this Wednesday as the sentencing date for a bin Laden associate for his role in the 1998 bombing of a U.S. embassy in Tanzania that killed 213 people. The sentencing had been set for the federal courthouse near the World Trade Center. No one from the U.S. attorney's office could be reached Tuesday to comment on whether the sentencing was still on.

Afghanistan's hardline Taliban rulers condemned the attacks and rejected suggestions that bin Laden was behind them, saying he does not have the means to carry out such well-orchestrated attacks. Bin

Like many of the workers, Diane wanted to do more. She wanted to help with the identifications. But the office had other people handling that job. One of them was New York City's only forensic anthropologist, Amy Mundorff, a young woman with long, dark curly hair. While working daily 12-hour shifts, Amy was recovering from cracked ribs, bruises, and a head injury. She had been standing near the south tower, responding to emergencies, when it collapsed.

Workers sift through debris from Ground Zero, looking for clues to solve the 9/11 crime and to identify human remains.

On Staten Island, no one knew if another attack would happen, but the possibility seemed very real. The bridge was closed. The Staten Island Ferry wasn't running. Security forces were checking everyone's identification.

At one point an urgent call came in over the radio. The trucks, conveyor belts, and other machinery drowned out most of it. The caller said: ". . . so don't worry. If we have to evacuate, we can get you out by helicopter." Later, Diane learned that the call was only a precaution. The United States had just begun bombing Afghanistan. Security forces were buttoning up New York City in case of a counterattack on U.S. soil.

Recovery teams raked the rubble, sometimes on their hands and knees, removing anything that looked like evidence.

Over the blaring noise, several of the rakers said to Diane, "We don't know if we're finding the right things."

Diane recognized the feeling: *Am I doing my best? Is there anything I can do better? How I can help?*

Diane replied, "You're doing a really great job." Everyone was. And Diane felt genuinely proud to be a part of it.

Anthropologists

study hundreds of skeletons but,

barring serious injury,

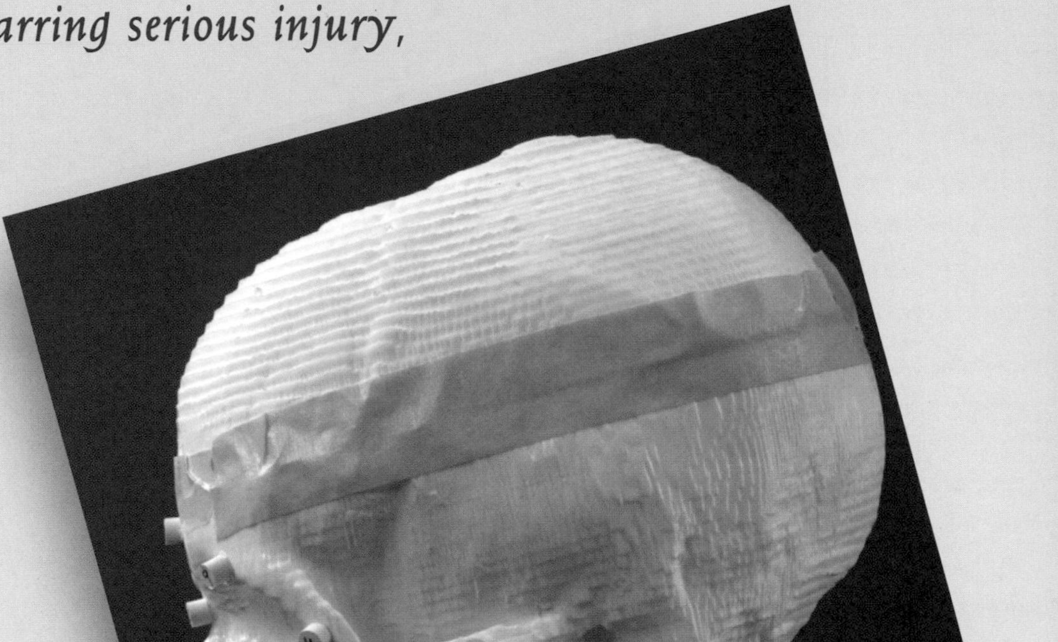

they never get to

see their own.

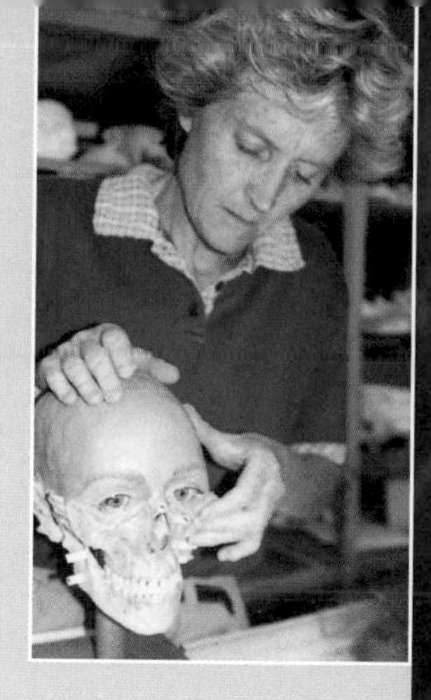

10

DIANE FRANCE'S SKULL?

On the wooden shelves of France Casting, one skull has a permanent perch amid the changing assortment of bones and casts. There's no label, so a curious visitor would have to guess its identity. If the visitor happens to be an anthropologist, the skull's secrets emerge quickly from the bone details:

A skull cast *(opposite)* is prepped to receive a sculpted face of clay. The numbered pegs show the sculptor how thick to make the "flesh." In the photo above, Diane removes a clay face to examine a murder victim's actual skull bones beneath.

The face is long and narrow, as is the nasal opening. The mouth isn't very wide. In profile, the contour of the jawline from ear to chin is softly curved. It's not at all a "Jay Leno" chin, but it ends in a distinct point. A careful measure of the cranium's height and width reveals that its shape is medium—neither long nor round. These and other features strongly suggest European ancestry.

The orbital ridges around the eye socket are sharp. There are no brow ridges above them. The muscle attachment points, such as the mastoid process, are small. All of these features are much more common in females than males.

The woman is full grown—since all her molars have grown in—but her exact age is inconclusive. It's impossible to determine without additional skeletal evidence.

So far this skull could belong to millions of people. The anthropologist looks for unique details to narrow the field.

On the inside of the frontal bone (forehead), just to the left of the midline of the face and near the hairline, is a tiny tumor.

The little tumor is benign—a harmless oddity that usually goes unnoticed throughout life. In anthropology, however, such oddities are excellent clues for identification. The anthropologist compares the skull to a Computed Tomography (CT) scan, a detailed computer image of the inside of a head, and finds the same tiny tumor. Matching the teeth with dental records would cement the conclusion: *This is the skull of Diane France.*

Reconstructing Diane

Facial reconstruction is the art and science of creating a face based on the shape of a skull. The scientific part of reconstruction works off the idea that a face is like a form-fitting mask over the skull. The shapes, sizes, and relative positions of the bone features match those of the face. For example, if the nasal opening of a skull is very narrow, then the nose is, too.

The artistic part of reconstruction is trickier because it's about filling in facial features that can't be determined by bone alone.

What would happen, then, if more than one anthropologist reconstructed the same skull? Would they produce the same face?

Anthropologist Norm Sauer set up an experiment to find out and enlisted Diane's help, or, more accurately, her skull. A CT scan of her head was turned into a 3-D model of her skull. From it Diane created identical plastic casts for each of six anthropologists. The scientists were told only that the subject was female, of European ancestry, and 30 to 50 years old. They then had to reconstruct th face.

Jennifer Fillion was one of the volunteer scie tists. Commenting on the experiment's artis aspect she explained, "We can't tell how fat face was or the shape of the cartilage portic of the nose—the middle to the tip."

Rather than guess, the reconstructionists st to a generalized nose. Diane's nose is a bit unusual—narrow and sharp. Jennifer's reco struction *(shown right)* depicts a more ordinary, rounder nose.

Jennifer added, "There are subtle variation in lip shape, ear shape, and the thickness a shape of the eyelids." Eyelids might be hea or sunken, for example. She doesn't add ey lashes or eyebrows, which vary a lot.

How much do reconstructions wind up look ing like the person? How useful are they in identifying unknown skeletons? It's an ongoing debate. Jennifer believes they're use if you ignore individual features—eyes, nos mouth. Look instead, she says, at "the over

~ Flesh and Bone

Diane picks up "her" skull and turns it around in her hands, examining the details. It isn't the skull in her head, of course. It's a plastic cast she made from a model. The model was used in a science experiment on the accuracy of facial reconstruction—sculpting a clay face on top of a skull. *(See box.)*

Diane displays a plastic cast of her own skull, made from a CT scan of her head. Without knowing it was Diane's, anthropologist Jennifer Fillion of Michigan State University used the cast to sculpt the face below. Do you see a likeness?

composition of the face and the combination of the facial features together."

Norm Sauer believes the greatest benefit of a reconstruction isn't identification; it's to draw attention to an unsolved murder case. He says people pay more attention to faces than to numbers such as height, weight, and age.

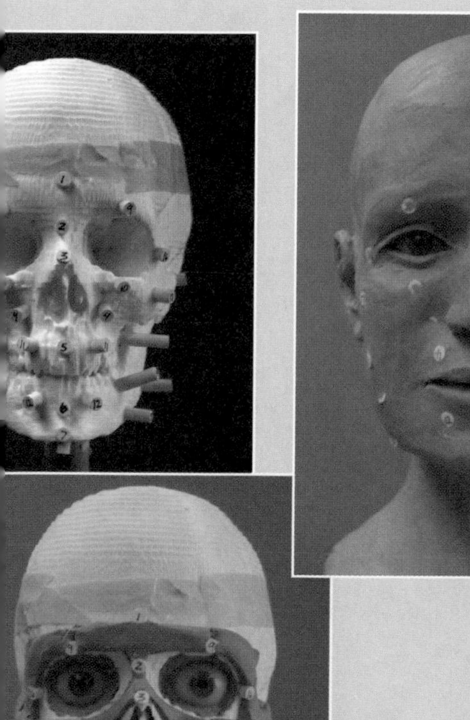

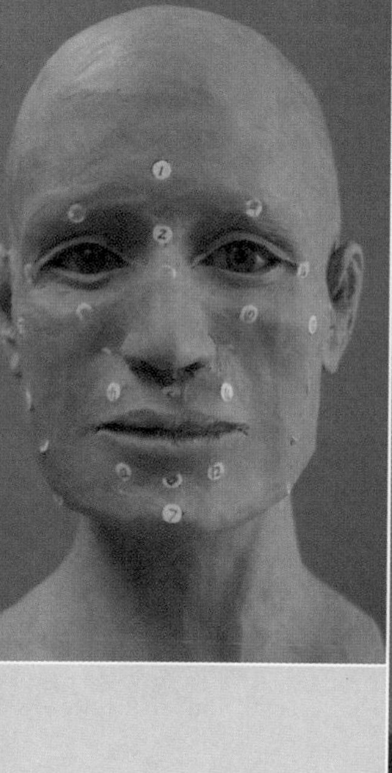

Anthropologists study hundreds of skeletons but, barring any serious injury, they never get to see their own. It's a ghoulish thought, but true. In fact, the only part of the skeleton that most people ever see is their teeth. Diane feels incredibly lucky to be looking at what lies beneath her living flesh.

She smiles as she recalls her father's reaction. Casually holding up the skull cast one day, she asked him, "Does this look familiar?"

Dad had no idea who it was. When she told him that it was her own skull, he was truly puzzled. He looked at his daughter's face, then at his daughter's skull, and then back at his daughter holding her own skull. . . . How was that possible?

The technology is borrowed from the automobile industry. Carmakers design new models using computer-aided design (CAD) programs. CAD allows users to look at an image from any angle—inside or out. A laser machine called a rapid prototyper builds a three-dimensional model of the CAD image out of plastic or other materials. Diane substituted a CT scan of her skull for a CAD design of a car, and now she's holding her head in her hands.

Moki *(left center)* Diane *(lower right, with Handy)*, and assorted friends get ready for a trail ride at Bonner Peak Ranch in Colorado.

Diane France replaces her skull on the shelf and walks barefoot into the molding area. It's 2004, and she turns 50 years old this year, in May. She now has two assistants in her casting lab, and the three of them have casting orders to fill. Sadly, her malamute friend, Moki, died just before last Christmas, at age 13. Diane is still feeling the loss. She has lived with a dog by her side since Ruby, the Norwegian elkhound she grew up with in Walden. Now, Diane is teaching her new puppy, an Australian shepherd named Lucy, how to hang out in the lab without poking her nose into things.

~ *The Adventures Continue*

Though the casting business is a success, Diane can't stop thinking and tinkering. She loves to try new formulas for plastic and to experiment with molding techniques. She never knows what—or who—her next cast might be.

The field of anthropology is growing, changing, and advancing, and Diane sees herself as a lifelong student as well as a teacher. She sometimes gets a chuckle when she passes out her casts in forensic anthropology classes at CSU. Asked to identify the ancestry of a bone, some students see "France" written on the back of the cast and write down "European." (Diane signs every cast by hand.) She makes sure her students leave her class knowing how to "read" bones scientifically.

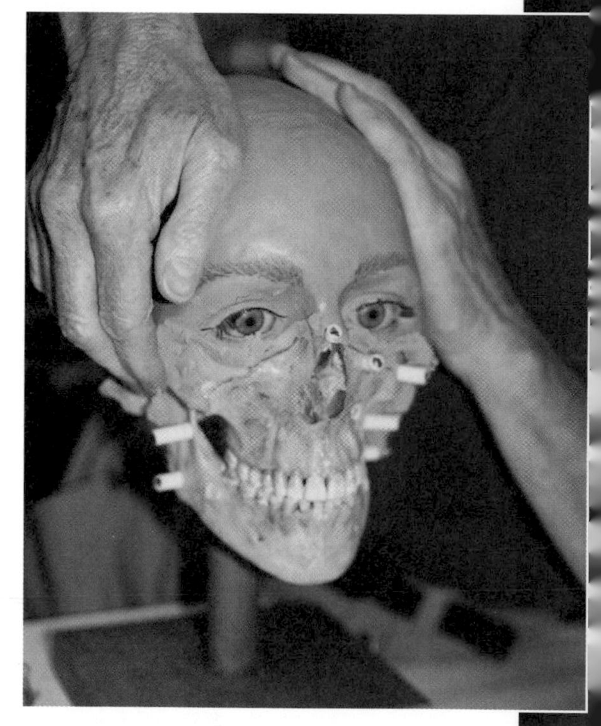

Who is this young woman? No one recognized the face sculpted directly onto her skull. Underneath the clay, Diane found bone evidence that she was murdered.

Fewer than 75 forensic anthropologists are certified by the American Board of Forensic Anthropology (ABFA), but the field is still young and growing. Diane's ABFA certificate is number 41. She knows that she has a rare and much-needed skill that can help grieving families, and she wants to use it. As the director of CSU's Human Identification Lab, she receives local murder victims and unidentified bodies to examine and identify. She's always on call for DMORT to fly anywhere in the world at any time.

Diane considers her friends at NecroSearch as family now. The group has 35 volunteer members and has worked more than 200 cases. Their work has taken them to 38 states and seven countries. They continue to search for clandestine graves, taking on about 20 cases per year. They also organize workshops to teach scientists and law enforcement officials how to search for clandestine graves.

Diane can't count the number of cases she's worked on, but on-the-scene details come back to her vividly. When a reporter asks her what her dream case would be, she doesn't hesitate: "To find Jimmy Hoffa, the missing person that everyone's looking for." The labor union leader disappeared mysteriously—most likely murdered—in Michigan in 1975. Tips on the location of his body still surface now and then.

Solving such a famous case would be a true adventure to Diane France. And, as she says, "If you have really good adventures, then you've had a good life."

Timeline of Diane France's Life

1954 Diane France is born on May 11.

1957 Mike France, her younger brother, is born.

1971 Diane spends the summer in Finland.

1972 She graduates second in her class from North Park High School in Walden, Colorado.

1976 Diane earns a bachelor's degree in anthropology from Colorado State University (CSU).

1979 Diane earns a master's degree in anthropology from CSU.

1980 Archaeology professor Cal Jennings and Diane France marry in a small family ceremony.

1983 Diane earns a Ph.D. in physical anthropology from CSU.

1985 On her first mass fatality case, Diane identifies victims of a gas explosion and fire in Glenwood Springs, Colorado.

1988 Diane joins NecroSearch, a new volunteer group that specializes in finding clandestine graves.

1989 The American Board of Forensic Anthropology certifies Diane as the 41st forensic anthropologist in the United States.

1992 NecroSearch finds the remains of Michele Wallace, who disappeared in 1974.

1994 Diane testifies at the trial of Roy Melanson, who is convicted of murdering Michele Wallace.

1995 The National Zoo's Think Tank exhibit opens, featuring Diane's bronze brain casts. Diane casts the skull of the man assumed to be Jesse James. NecroSearch finds the body of Cher Elder, murdered in 1993.

1996 Tom Luther is convicted of the murder of Cher Elder.

1997 Diane casts the tongue and paw of a living tiger for a National Zoo exhibit. As a member of DMORT, she identifies victims of a Korean Air crash in Guam.

1998 Diane travels to Russia to investigate the case of the two missing Romanov children. She and Cal divorce.

1999 Diane joins a DMORT team at the recovery of an Egypt Air crash. She is diagnosed with thyroid cancer.

2000 NecroSearch elects Diane as president. Diane is treated for thyroid cancer and has a full recovery.

2001 After the terrorist attack of September 11, Diane joins a DMORT team in New York City.

2004 After completing casts of "the *Hunley* boys," Diane attends events in honor of the eight Confederate soldiers who drowned when the submarine sank.

2005 In July, Diane marries Arthur Abplanalp, a long-time friend of the France family. Diane continues to teach forensic identification at the AFIP. NecroSearch plans a class for investigators worldwide involved in human rights abuse cases and mass graves.

About the Author

Lorraine Jean Hopping learned to love science by writing about it. She was the founding editor of Scholastic Inc.'s *SuperScience* magazine for students in grades four to six and has written 30 books for children of all ages, including another biography in this series, *Space Rocks*. She also invented the award-winning Mars 2020 and a dozen other board games published by Aristoplay. Her latest adventure in science is editing Joy Hakim's *Story of Science* book series. Lorraine lives in Ann Arbor, Michigan, with her husband Chris and two cats.

Glossary

This book is about bones. The word bone, so short and simple, comes from Old English. But if you want to speak scientifically, you'll need a little Greek and Latin. Long science words are easy to figure out if you look for patterns: *osteology* (the study of bones), *osteoblasts* (bone-building cells), *osteometric* board (bone-measuring tool), *osseous* (bony), *ossify* (turn to bone). . . . Guess what? The Greek word for bone is *ostéon*; the Latin word is *ossis* (or *os*). If you see either root in a word, there's an excellent chance that the word has to do with bones. This list will help you "bone up" on other science words in this book. For more information about each word, use your dictionary.

anatomy: the structure of a plant or animal. When a word ends in *–tomy*, think "cut." For centuries, scientists have cut up, or dissected, bodies to learn about them. The word *anatomy* used to mean "dissection," but now it refers to the science of the body in general. The adjective is *anatomical.*

anthropology: the study of humans, especially their physical remains, artifacts, and cultures

archaeology: the scientific study of the peoples, cultures, and life of ancient times

artifact: anything made by humans, especially tools or weapons

cadaver: a dead body or corpse (*corpus* means "body") used for science and medical studies

cartilage: the gristle between bones—that white, smooth, crunchy stuff at the end of a chicken leg, for example. It keeps bones from grinding against each other at the joints.

coroner: the person who investigates the medical aspects of a death. A coroner examines the body, inside and out, in an autopsy, for example, to identify cause of death.

cranium: the part of the skull enclosing the brain

criminalist: an expert in the scientific study of crimes. A criminalist tests blood for DNA, matches bullets with guns, lifts fingerprints, and so on. A crime scene investigator is a criminalist who examines and collects evidence at a crime scene.

curator: a person in charge of all or a section of a museum, zoo, or art gallery

decompose: to rot, decay, or break down. The process is called *decomposition*. Organisms that help that process along are called *decomposers*.

DNA (deoxyribonucleic acid): a nucleic acid in the chromosomes of living cells that carries the genetic code

femur: the thigh bone

forensic: used in a court of law or public debate. Forensic anthropology is the science of processing physical evidence such as human remains in an investigation. The word is related to the Latin *forum*—a public meeting place.

humerus: the long bone in the upper part of the arm from the shoulder to the elbow

ligament: connective fibers that attach one bone to another. In Latin, it means a "tie or binding."

odontologist: a branch of anatomy dealing with the structure of teeth

specimen: any sample—animal, vegetable, or mineral—used for scientific study. It could be a bear claw, a flower, a rock, or, at the National Museum of Health and Medicine, a body part preserved in a jar.

sternum: the breastbone

tendon: a strong band that connects a muscle to a bone

tibia: the inner and thicker of the two bones from the knee to the ankle; the shinbone.

vertebrate: any animal that has a spinal column. The individual bones of the backbone are *vertebrae*.

Metric Conversion Chart

When you know:	Multiply by:	To convert to:
Inches	2.54	Centimeters
Feet	0.30	Meters
Pounds	0.45	Kilograms
Gallons	3.79	Liters
Centimeters	0.39	Inches
Meters	3.28	Feet
Kilograms	2.2	Pounds
Liters	0.26	Gallons

Further Resources

Women's Adventures in Science on the Web

Now that you've met Diane France and learned all about her work, are you wondering what it would be like to be a forensic anthropologist? How about an astronomer, a wildlife biologist, or a robot designer? It's easy to find out. Just visit the *Women's Adventures in Science* Web site at **www.iWASwondering.org.** There you can live your own exciting science adventure. Play games, enjoy comics, and practice being a scientist. While you're having fun, you'll also get to meet amazing women scientists who are changing our world.

BOOKS

Colman, Penny. *Corpses, Coffins and Crypts: A History of Burial.* New York: Henry Holt, 1997. Death isn't everyone's favorite subject, but it can teach us a lot about society and culture. This book lets you tackle this tough topic head-on and, surprisingly, with a little humor. It's part anthropology, part history and includes details and photographs of decomposition, graves, and burial practices.

McGowan, Christopher. *Make Your Own Dinosaur Out of Chicken Bones.* New York: Harper Collins, 1997. This isn't one of those kits with plastic pieces that you snap together. It's the real thing. Learn how to take clean chicken skeletons left over from dinner and arrange the loose bones into a dinosaur model. This works because vertebrates (chickens, dinosaurs, you) have the same basic bones.

Owen, David. *Police Lab: How Forensic Science Tracks Down and Convicts Criminals.* Buffalo, New York: Firefly Books (U.S.) Inc., 2002. Read about forensic scientists in action. Twenty real-life cases show how forensic scientists solve crimes through crime-scene analysis of weapons and physical clues, DNA testing, and facial reconstruction.

WEB SITES

France Casting: www.francecasts.com
Take a look at Diane's realistic-looking casts of humans, other primates, and nonhuman animals.

The Hunley: www.history.navy.mil and www.hunley.org
The Naval Historical Center of the U.S. Navy has interesting diagrams, photos, and documents. The Friends of the Hunley is the organization that raised and excavated the submarine and created a museum for it in Charleston, South Carolina.

The National Zoo: www.nationalzoo.si.edu/Animals
Diane's bronze brain casts are featured in the "Think Tank" exhibit. You can also learn more about the zoo's long-tongued tigers by visiting the "Great Cats" page.

NecroSearch International: www.necrosearch.org
Meet the team that searches high and low (mostly low) for clandestine graves. Find out what they do, and a little about how they do it.

Selected Bibliography

In addition to interviews with Diane France, her family, and her friends, the author did extensive reading and research to write this book. Here are some of the sources she consulted.

Alexander, R. McNeill et al. *Bones: The Unity of Form and Function.* New York: MacMillan, 1994.

Baun, Paul. *Written in Bones: How Human Remains Unlock the Secrets of the Dead.* Buffalo, New York: Firefly Books (U.S.) Inc., 2002.

France, Diane L. *Lab Manual and Workbook for Physical Anthropology.* 5th edition. Belmont, California: Wadsworth/Thomson Learning, 2004.

Hoyt, Edwin Palmer. *The Voyage of the Hunley.* Springfield, New Jersey: Burford Books, 2002.

Jackson, Steve. *No Stone Unturned: The Story of NecroSearch International Investigators.* New York: Pinnacle Books, 2002.

Maples, William R. and Michael Browning. *Dead Men Do Tell Tales.* New York: Doubleday, 1994.

Massie, Robert K. *The Romanovs: The Final Chapter.* New York: Random House, 1995.

Ubelaker, Douglas H. and Henry Scammell. *Bones: A Forensic Detective's Casebook.* New York: Harpercollins, 1992.

INDEX

WOMEN'S ADVENTURES IN SCIENCE ADVISORY BOARD

A group of outstanding individuals volunteered their time and expertise to help the Joseph Henry Press conceptualize and develop the *Women's Adventures in Science* series. We thank the following people for their generous contribution of time and talent:

Maxine Singer, Advisory Board Chair: Biochemist and former President of the Carnegie Institution of Washington

Sara Lee Schupf: Namesake for the Sara Lee Corporation and dedicated advocate of women in science

Bruce Alberts: Cell and molecular biologist, President of the National Academy of Sciences and Chair of the National Research Council, 1993–2005

May Berenbaum: Entomologist and Head of the Department of Entomology at the University of Illinois, Urbana-Champaign

Rita R. Colwell: Microbiologist, Distinguished University Professor at the University of Maryland and Johns Hopkins University, and former Director of the National Science Foundation

Krishna Foster: Assistant Professor in the Department of Chemistry and Biochemistry at California State University, Los Angeles

Alan J. Friedman: Physicist and Director of the New York Hall of Science

Toby Horn: Biologist and Co-director of the Carnegie Academy for Science Education at the Carnegie Institution of Washington

Shirley Jackson: Physicist and President of Rensselaer Polytechnic Institute

Jane Butler Kahle: Biologist and Condit Professor of Science Education at Miami University, Oxford, Ohio

Barb Langridge: Howard County Library Children's Specialist and WBAL-TV children's book critic

Jane Lubchenco: Professor of Marine Biology and Zoology at Oregon State University and President of the International Council for Science

Prema Mathai-Davis: Former CEO of the YWCA of the U.S.A.

Marcia McNutt: Geophysicist and CEO of Monterey Bay Aquarium Research Institute

Pat Scales: Director of Library Services at the South Carolina Governor's School for the Arts and Humanities

Susan Solomon: Atmospheric chemist and Senior Scientist at the Aeronomy Laboratory of the National Oceanic and Atmospheric Administration

Shirley Tilghman: Molecular biologist and President of Princeton University

Gerry Wheeler: Physicist and Executive Director of the National Science Teachers Association

Library Advisory Board

A number of school and public librarians from across the United States kindly reviewed sample designs and text, answered queries about the format of the books, and offered expert advice throughout the book development process. The Joseph Henry Press thanks the following people for their help:

Barry M. Bishop
Director of Library Information Services
Spring Branch Independent School District
Houston, Texas

Danita Eastman
Children's Book Evaluator
County of Los Angeles Public Library
Downey, California

Martha Edmundson
Library Services Coordinator
Denton Public Library
Denton, Texas

Darcy Fair
Children's Services Manager
Yardley-Makefield Branch
Bucks County Free Library
Yardley, Pennsylvania

Kathleen Hanley
School Media Specialist
Commack Road Elementary
Islip, New York

Amy Louttit Johnson
Library Program Specialist
State Library and Archives of Florida
Tallahassee, Florida

Mary Stanton
Juvenile Specialist
Office of Material Selection
Houston Public Library
Houston, Texas

Brenda G. Toole
Supervisor, Instructional Media Services
Panama City, Florida

Student Advisory Board

The Joseph Henry Press thanks students at the following schools and organizations for their help in critiquing and evaluating the concept for the book series. Their feedback about the design and storytelling was immensely influential in the development of this project.

The Agnes Irwin School, Rosemont, Pennsylvania
La Colina Junior High School, Santa Barbara, California
The Hockaday School, Dallas, Texas
Girl Scouts of Central Maryland, Junior Girl Scout Troop #545
Girl Scouts of Central Maryland, Junior Girl Scout Troop #212

JHP Executive Editor: Stephen Mautner

Series Managing Editor: Terrell D. Smith

Designer: Francesca Moghari

Illustration research: Joan Mathys

Special contributors: Meredith DeSousa, Allan Fallow, Mary Kalamaras, April Luehmann, Mary Beth Oelkers-Keegan, Moses Schanfield, Anita Schwartz

Graphic design assistance: Michael Dudzik

Illustration Credits:

Cover, viii, x Joshua Buck; **1** Paul Sledzik, National Museum of Health and Medicine, Armed Forces Institute of Pathology; **2** *(t)* Courtesy Tom Adair and NecroSearch International; **2** *(b)*, **3, 4** Paul Sledzik, National Museum of Health and Medicine, Armed Forces Institute of Pathology; **5** Diane France; **8** Courtesy Terrell Smith; **9** Jessie Cohen, Smithsonian National Zoo; **10, 11, 12, 13** *(m, b)*, **14** *(both)* Courtesy Diane France; **15** Courtesy Joy Metzger; **18, 19, 20** Courtesy Diane France; **22** Courtesy of University Archives, The Bancroft Library, University of California, Berkeley; **23** Diane France; **26** Corbis/Royalty Free; **28** *(t)* Paul Sledzik, National Museum of Health and Medicine, Armed Forces Institute of Pathology; *(b)* National Museum of Health and Medicine, Armed Forces Institute of Pathology; **34** Courtesy Diane France; *(inset)* Bryn Mawr College, Marian Coffin Canaday Library; **36** Diane France; **37** Reprinted with permission of the *Rocky Mountain News*, photographers George Kochanec and Doug Stewart, 12/17/85; **38** Courtesy Tom Adair and NecroSearch International; **39** *(both)* Reprinted with permission of the *Rocky Mountain News*, photographers George Kochanec and Doug Stewart, 12/17/85; **40, 41** Courtesy Diane France; **44, 45, 47, 48, 49** Courtesy Tom Adair and NecroSearch International; **51** Courtesy Jim Reed and NecroSearch International; **52** Courtesy Gunnison County, Colorado, Sheriff's Office; **53** Courtesy Tom Adair and NecroSearch International; **54, 55, 56, 58** Courtesy Gunnison County, Colorado, Sheriff's Office; **59, 60** *(both)* Courtesy Lakewood, Colorado, Police Department; **62** Library of Congress; **63** © Agence No 7/Corbis Sygma; **65** *(both)* Sovfoto; **66-67, 68** *(t)* Courtesy Jim Reed; *(b)* East News; **69** Courtesy Diane France; **70** © Rykoff Collection/ Corbis; **71** Sovfoto; **74** East News; **76** East News; *(inset)* © V.Velengurin/R.P.G./Corbis Sygma; **78** Library of Congress; **79** Courtesy Diane France; **80** Library of Congress; **81** State Historical Society of Missouri, Columbia; **82** Diane France; **83** *(t)* U.S. Naval Historical Center Photograph; *(b)* Museum of the Confederacy, Richmond, Virginia, photograph by Katherine Wetzel; **84** Barbara Vulgaris, U.S. Naval Historical Center; **85** *(all)* Courtesy Friends of the Hunley; **87** *(t)* Mary Wright; *(b)* Diane France; **88** *(t)* Diane France; *(b)* Jessie Cohen, Smithsonian National Zoo; **90** U.S. Coast Guard; **91** Courtesy Jim Reed and NecroSearch International; **92** Paul Sledzik; **95** Heywood Sawyer; **96** Fort Collins *Coloradoan*; **97** AP/Wide World Photos; **98** Courtesy Jennifer Fillion, Michigan State University Forensic Anthropology Laboratory; **99** Lorraine Hopping Egan; **101** *(t)* Lorraine Hopping Egan; *(b)* Courtesy Jennifer Fillion, Michigan State University Forensic Anthropology Laboratory; **102** Heywood Sawyer; **103** Lorraine Hopping Egan; **104** Diane France.

Maps: © 2004 Map Resources
Illustrations: Kathryn Born

The border image used throughout the book is a detail of a radiograph of a spine (Corbis Royalty-Free).